TRANSMISSION LINES IN DIGITAL SYSTEMS FOR EMC PRACTITIONERS

TRANSMISSION LINES IN DIGITAL SYSTEMS FOR EMC PRACTITIONERS

CLAYTON R. PAUL

Department of Electrical and Computer Engineering
Mercer University
Macon, Georgia
and
Emeritus Professor of Electrical Engineering
University of Kentucky
Lexington, Kentucky

IEEE PRESS

A JOHN WILEY & SONS, INC., PUBLICATION

Published by John Wiley & Sons, Inc., Hoboken, New Jersey.
Published simultaneously in Canada.

For general information on our other products and services or for technical support, please contact our Customer Care Department within the United States at (800) 762-2974, outside the United States at (317) 572-3993 or fax (317) 572-4002.

Wiley also publishes its books in a variety of electronic formats. Some content that appears in print may not be available in electronic formats. For more information about Wiley products, visit our web site at www.wiley.com.

Library of Congress Cataloging-in-Publication Data:

Paul, Clayton R.
  Transmission lines in digital systems for EMC practitioners / Clayton R. Paul.
     p. cm.
   Includes bibliographical references and index.
   ISBN 978-1-118-14399-5 (hardback)
1. Electromagnetic compatibility. 2. Telecommunication lines. I. Title.
   TK7867.2.P383. 2012
   621.382′24–dc23
                                                       2011021000

oBook ISBN: 9781118145579
ePDF ISBN: 9781118145548
ePub ISBN: 9781118145562
MOBI ISBN: 9781118145555

10 9 8 7 6 5 4 3 2 1

*This book is dedicated to the humane and compassionate treatment
of animals*

and my beloved pets:

*Patsy, Dusty, Megan, Tinker, Bunny, Winston, Sweetheart, Lady, Tigger,
Beaver, Ditso, Buru, Old Dog, Zip, Tara, Timothy, Kiko, Valerie, Red, Sunny,
Johnny, Millie, Molly, Angel, Autumn, and Shabby.*

*Those readers who are interested in the humane and compassionate
treatment of animals are encouraged to donate to*

*The Clayton and Carol Paul Fund for Animal Welfare
c/o the Community Foundation of Central Georgia
277 MLK, Jr. Blvd.
Suite 303
Macon, GA 31202*

The primary and only objective of this Fund is to provide monetary grants to

(1) animal humane societies
(2) animal shelters
(3) animal adoption agencies
(4) low-cost spay-neuter clinics
(5) individual wildlife rehabilitators
(6) as well as other organizations devoted to animal welfare

in order to allow these volunteer organizations to use their enormous
enthusiasm, drive and willingness to reduce animal suffering and home-
lessness through the monetary maintenance of their organizations where
little or no monetary funds existed previously.

CONTENTS

PREFACE

Most of the numerous textbooks I have published were intended for class instructional books for electrical engineering (EE) and computer engineering (CpE) courses in a university environment. I decided to write this book for the industrial professional. My work has been in the field of electromagnetic compatibility (EMC), known more commonly as interference in electronic systems. In the course of my teaching, I have also had the pleasure of working with many EMC professionals. It is to this group of professionals that I have focused the book.

I have written a brief but comprehensive book covering the set of *transmission-line* skills that EMC practitioners today require in order to be successful in high-speed digital electronics. The basic skills in the book weren't studied in most curricula some ten years ago. The rapidly changing digital technology has created this demand for a discussion of new analysis skills, particularly for the analysis of *transmission lines* where the conductors that interconnect the electronic modules have become "electrically large," longer than a tenth of a wavelength, which are becoming increasingly important. *Crosstalk* between the lines is also rapidly becoming a significant problem in getting modern electronic systems to work satisfactorily. Hence this small volume is concentrated on modeling "electrically long" connection conductors where previously used Kirchhoff's voltage and current laws and lumped-circuit modeling have become obsolete because of the increasing speeds of modern digital systems. One important exception is Chapter 5, where electrically short lines are considered exclusively when we consider the use of shielded lines and twisted pairs of wires to eliminate or reduce crosstalk for electrically short lines.

Until as recently as some ten years ago, digital system clock speeds and data rates were in the low megahertz range. The "lands" on printed circuit boards (PCBs) that interconnect the electronic modules had little or no impact on the proper functioning of those electronic circuits. Today, the clock and data speeds have moved into the low gigahertz range. As the demand for faster data processing continues to escalate, these speeds will no doubt continue to increase into the gigahertz frequency range. In addition, analog communication frequencies have also moved steadily into the gigahertz range and will no doubt continue to increase. Although the "physical dimensions" of these lands and the PCBs supporting them have not changed significantly over the intervening years, the spectral content of the signals they carry has increased significantly. Because of this, the "electrical dimensions" (in wavelengths) of the lands have increased to the point where these "interconnects" have a significant effect on the signals they are carrying, so that just getting the systems to work properly has become a major design problem. Prior to some ten years ago, these interconnects could be reliably modeled with lumped-circuit models that are easily analyzed using Kirchhoff's voltage and current laws and other lumped-circuit analysis methods. Because these interconnects are becoming "electrically long," lumped-circuit modeling of them is becoming inadequate and gives erroneous answers. Most of the interconnect conductors must now be treated as distributed-circuit *transmission lines*.

In Chapter 1, the increasingly important fundamental concepts of waves, wavelength, time delay, and electrical dimensions are discussed. In addition, the bandwidth of digital signals and its relation to pulse rise and fall times are discussed. The effect of electrically long conductors on signal integrity is discussed.

Chapter 2 covers the time-domain analysis of two-conductor transmission lines. The transmission-line equations are derived and solved, and the important concept of characteristic impedance is covered. The important per-unit-length parameters of inductance and capacitance that distinguish one line from another are obtained for typical lines. The terminal voltages and currents of lines with various source waveforms and resistive terminations are computed by hand via wave tracing. This gives considerable insight into the general behavior of transmission lines in terms of forward- and backward-traveling waves and their reflection. The SPICE computer program and its personal computer version, PSPICE, contains an *exact* model for a two-conductor lossless line and is discussed as a computational aid in solving for the transmission-line terminal voltages and currents. SPICE is an important computational tool, since it provides a determination of the terminal voltages and currents for practical linear and nonlinear terminations such as CMOS and bipolar devices, for which hand analysis is very formidable. Matching schemes for achieving signal integrity are covered, as are the effects of line discontinuities. Chapter 3 covers the corresponding analysis in the frequency domain. The important analog concepts of input impedance to the line and high-frequency modeling of electronic circuits are also discussed.

The remaining chapters, Chapters 4, 5, and 6, cover crosstalk between adjacent transmission lines. Chapter 4 covers the derivation of the multi-conductor transmission line (MTL) equations, consisting of lines having three conductors and the crosstalk between these lines. The derivation of the per-unit-length parameters in the 2×2 inductance and capacitance matrices, **L** and **C,** for three-conductor lines are discussed either in an approximation fashion for widely spaced lines or in terms of computer programs for handling these difficult calculations. These computer programs can be downloaded from the John Wiley ftp site:

ftp://ftp.wiley.com/public/sci_tech_med/multiconductor_transmission/

Their use is discussed for various cross-sectional geometries.

Chapter 5 discusses an approximate solution of the MTL equations for electrically short lines. This model is not only easy to use to compute the crosstalk in an approximate fashion but also illustrates the use of (1) shielded wires and (2) twisted pairs of wires for eliminating crosstalk. Numerous experimental results are shown that verify the model. The effect of pigtails in degrading the effect of shielded wires is shown along with experimental results.

Finally, Chapter 6 discusses an exact PSPICE subcircuit for the solution of any lossless MTL. Several experimental results that verify the accuracy of this model are shown. A computer program, SPICEMTL, that generates a PSPICE subcircuit model that can readily be embedded in a SPICE program to provide the *exact* solution can be downloaded from the ftp site.

Several important features of this textbook are (1) the basics of transmission-line fundamentals, (2) the numerous experimental results that illustrate and verify the mathematical results, and (3) the availability of computer programs that facilitate the solution of transmission lines for determining signal integrity as well as crosstalk.

The appendix gives a brief tutorial of SPICE (PSPICE), which is used extensively throughout the book.

CLAYTON R. PAUL

Macon, Georgia

1

TRANSMISSION LINES: PHYSICAL DIMENSIONS VS. ELECTRIC DIMENSIONS

With the operating frequencies of today's high-speed digital and high-frequency analog systems continuing to increase into the GHz ($1\,\mathrm{GHz} = 10^9\,\mathrm{Hz}$) range, previously-used lumped-circuit analysis methods such as Kirchhoff's laws *will no longer be valid* and will give *incorrect answers*. Physical dimensions of the system that are "electrically large" (greater than a tenth of a wavelength) *must be analyzed* using the transmission-line model. The wavelength, λ, of a single-frequency sinusoidal current or voltage wave is defined as $\lambda = v/f$ where v is the velocity of propagation of the wave on the system's conductors, and f is the cyclic frequency of the single-frequency sinusoidal wave on the conductor. Velocities of propagation on printed circuit boards (PCBs) lie between 60 and 100% of the speed of light in a vacuum, $v_0 = 2.99792458 \times 10^8\,\mathrm{m/s}$. A 1-GHz single-frequency sinusoidal wave on a pair of conductors of total length $\mathscr{L}$ will be one wavelength for

$$\mathscr{L} = 1\lambda = \frac{v_0 = 3 \times 10^8}{f = 1 \times 10^9} = 30\,\mathrm{cm} \cong 11.8\,\mathrm{in}$$

In this case the largest circuit that can be analyzed successfully using Kirchhoff's laws and lumped-circuit models is of length $\mathscr{L} = \frac{1}{10}\lambda = 3\,\mathrm{cm}$

Transmission Lines in Digital Systems for EMC Practitioners, First Edition.
Clayton R. Paul.
© 2012 John Wiley & Sons, Inc. Published 2012 by John Wiley & Sons, Inc.

= 1.18 in! In that case "electrically long" pairs of interconnect conductors ($\mathscr{L} > 3$ cm $= 1.18$ in) that interconnect the electronic modules must be treated as *transmission lines* in order to give correct answers.

The spectral (frequency) content of modern high-speed digital waveforms today as well as the operating frequencies of analog systems extend into the gigahertz regime. A digital clock waveform has a *repetitive* trapezoidal shape, as illustrated in Fig. 1.1. The period T of the periodic digital waveform is the reciprocal of the clock fundamental frequency, f_0, and the fundamental radian frequency is $\omega_0 = 2\pi f_0$. The rise and fall times are denoted τ_r and τ_f, respectively, and the pulse width (between 50% levels) is denoted τ. As the fundamental frequencies of the clocks, f_0, are increased, their period $T = 1/f_0$ decreases and hence the rise and fall times of the pulses must be reduced commensurately in order that pulses resemble a trapezoidal shape rather than a "sawtooth" waveform, thereby giving adequate "setup" and "hold" time intervals. Typically, the rise and fall times are chosen to be 10% of the period T to achieve this. Reducing the pulse rise and fall times has had the consequence of increasing the spectral content of the waveshape. Typically, this spectral content is significant up to the inverse of the rise and fall times, $1/\tau_r$. For example, a 1-GHz digital clock signal having rise and fall times of 100 ps (1 ps $= 10^{-12}$ s) has significant spectral content at multiples (harmonics) of the basic clock frequency (1 GHz, 2 GHz, 3 GHz, ...) up to around 10 GHz. Since the digital clock waveform shown in Fig. 1.1 is a periodic, repetitive waveform, according to the Fourier series their time-domain waveforms can be viewed alternatively as being composed of an infinite number of harmonically related sinusoidal components as

$$x(t) = c_0 + c_1 \cos(\omega_0 t + \theta_1) + c_2 \cos(2\omega_0 t + \theta_2) + c_3 \cos(3\omega_0 t + \theta_3) + \cdots$$
$$= c_0 + \sum_{n=1}^{\infty} c_n \cos(n\omega_0 t + \theta_n)$$

(1.1a)

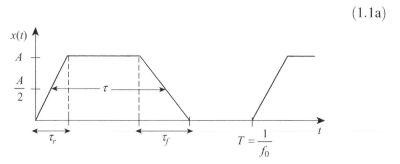

FIGURE 1.1. Typical digital clock/data waveform.

The constant component c_0 is the average (dc) value of the waveform over one period of the waveform,

$$c_0 = \frac{1}{T} \int_{t_1}^{t_1+T} x(t)\, dt \qquad (1.1b)$$

and the other coefficients (magnitude and angle) are obtained from

$$c_n \angle \theta_n = \frac{2}{T} \int_{t_1}^{t_1+T} x(t) e^{-jn\omega_0 t}\, dt \qquad (1.1c)$$

where $j = \sqrt{-1}$ and the exponential is complex-valued with a unit magnitude and an angle as $e^{-j\omega_0 t} = 1 \angle -\omega_0 t$.

In the past, clock speeds and data rates of digital systems were in the low megahertz ($1\,\text{MHz} = 10^6\,\text{Hz}$) range, with rise and fall times of the pulses in the nanosecond ($1\,\text{ns} = 10^{-9}\text{s}$) range. Prior to that time, the "lands" (conductors of rectangular cross section) that interconnect the electronic modules on PCBs were "electrically short" and had little effect on the proper functioning of those electronic circuits. The time delays through the modules dominated the time delay imposed by the interconnect conductors. Today, the clock and data speeds have moved rapidly into the low gigahertz range. The rise and fall times of those digital waveforms have decreased into the picosecond ($1\,\text{ps} = 10^{-12}\,\text{s}$) range. The delays caused by the interconnects have become the dominant factor.

Although the "physical lengths" of the lands that interconnect the electronic modules on the PCBs have not changed significantly over these intervening years, their "electrical lengths" (in wavelengths) have increased dramatically because of the increased spectral content of the signals that the lands carry. Today these "interconnects" can have a significant effect on the signals they are carrying, so that just getting the systems to work properly has become a major design problem. Remember that it does no good to write sophisticated software if the hardware cannot execute those instructions faithfully. This has generated a new design problem referred to as *signal integrity*. Good signal integrity means that the interconnect conductors (the lands) should not adversely affect operation of the modules that the conductors interconnect. Because these interconnects are becoming "electrically long," lumped-circuit modeling of them is becoming inadequate and gives erroneous answers. Many of the interconnect conductors must now be treated as distributed-circuit *transmission lines*.

1.1   WAVES, TIME DELAY, PHASE SHIFT, WAVELENGTH, AND ELECTRICAL DIMENSIONS

In the analysis of electric circuits using Kirchhoff's voltage and current laws and lumped-circuit models, we *ignored* the connection leads attached to the lumped elements. When is this permissible? Consider the lumped-circuit element having attachment leads of total length $\mathscr{L}$ shown in Fig. 1.2. *Single-frequency sinusoidal* currents along the attachment leads are, in fact, *traveling waves*, which can be written in terms of position z along the leads and time t as

$$\boxed{i(t,z) = I\cos(\omega t - \beta z)} \tag{1.2}$$

where the radian frequency ω is written in terms of cyclic frequency f as $\omega = 2\pi f$ rad/s and β is the *phase constant* in units of rad/m. (Note that the argument of the cosine must be in radians and not degrees.) To observe the movement of these current waves along the connection leads, we observe and track the movement of a point on the wave in the same way that we observe the movement of an ocean wave at the seashore. Hence the argument of the cosine in (1.2) must remain constant in order to track the movement of a point on the wave so that $\omega t - \beta z = C$, where C is a constant. Rearranging this as $z = (\omega/\beta)t - C/\beta$ and differentiating with

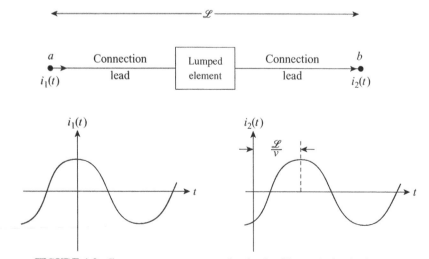

FIGURE 1.2. Current waves on connection leads of lumped-circuit elements.

respect to time gives the velocity of propagation of the wave as

$$v = \frac{\omega}{\beta} \qquad \frac{m}{s}$$

(1.3)

Since the argument of the cosine, $\omega t - \beta z$, in (1.2) must remain a constant in order to track the movement of a point on the wave, as time t increases, so must the position z. Hence the form of the current wave in (1.2) is said to be a *forward-traveling wave*, since it must be traveling in the $+z$ direction in order to keep the argument of the cosine constant for increasing time. Similarly, a *backward-traveling wave* traveling in the $-z$ direction would be of the form $i(t, z) = I \cos(\omega t + \beta z)$, since as time t increases, position z must decrease to keep the argument of the cosine constant and thereby track the movement of a point on the waveform. Since the current is a *traveling wave*, the current entering the leads, $i_1(t)$, and the current exiting the leads, $i_2(t)$, are separated in time by a *time delay* of

$$T_D = \frac{\mathscr{L}}{v} \qquad s$$

(1.4)

as illustrated in Fig. 1.2. These single-frequency waves suffer a *phase shift* of $\phi = \beta z$ radians as they propagate along the leads. Substituting (1.3) for $\beta = \omega/v$ into the equation of the wave in (1.2) gives an equivalent form of the wave as

$$i(t, z) = I \cos\left[\omega\left(t - \frac{z}{v}\right)\right]$$

(1.5)

which indicates that *phase shift is equivalent to a time delay*. Figure 1.2 plots the current waves *versus time*. Figure 1.3 plots the current wave *versus position in space at fixed times*.

As we will see, the critical property of a traveling wave is its wavelength, denoted λ. *A wavelength is the distance the wave must travel in order to shift its phase by 2π radians or $360°$*. Hence $\beta\lambda = 2\pi$ or

$$\lambda = \frac{2\pi}{\beta} \qquad m$$

(1.6)

Alternatively, the wavelength is the distance between the same adjacent points on the wave: for example, between adjacent wave crests, as illustrated in

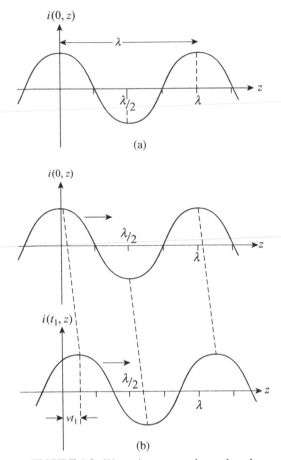

FIGURE 1.3. Waves in space and wavelength.

Fig. 1.3. Substituting the result in (1.3) for β in terms of the wave velocity of propagation v gives an alternative result for computing the wavelength:

$$\boxed{\lambda = \frac{v}{f} \quad \text{m}}$$ (1.7)

Table 1.1 gives the wavelengths of single-frequency sinusoidal waves in free space (essentially, air) where $v_0 \cong 3 \times 10^8\,\text{m/s}$. (The velocities of propagation of current waves on the lands of a PCB are less than in free space, which is due to the interaction of the electric fields with the board material. Hence wavelengths on a PCB are shorter than they are in free space.) Observe that a wave of frequency 300 MHz has a wavelength of 1 m. Note that the product of the frequency of the wave and its wavelength equals the velocity of

TABLE 1.1. Frequencies of Sinusoidal Waves in Free Space (Air) and Their Corresponding Wavelengths

Frequency, f	Wavelength, λ
60 Hz	3107 mil (5000 km)
3 kHz	100 km
30 kHz	10 km
300 kHz	1 km
3 MHz	100 m ($\approx$ 300 f)
30 MHz	10 m
300 MHz	**1 m ($\approx$ 3 f)**
3 GHz	10 cm ($\approx$ 4 in)
30 GHz	1 cm
300 GHz	0.1 cm

propagation of the wave, $f\lambda = v$. Wavelengths scale linearly with frequency. As frequency decreases, the wavelength increases, and vice versa. For example, the wavelength of a 7-MHz wave is easily computed as

$$\lambda|_{@7 \text{ MHz}} = \frac{300 \text{ MHz}}{7 \text{ MHz}} \times 1 \text{ m} = 42.86 \text{ m}$$

Similarly, the wavelength of a 2-GHz cell phone wave is 15 cm, which is approximately 6 in.

Now we turn to the important criterion of physical dimensions in terms of wavelengths: that is, "electrical dimensions." To determine a physical dimension, $\mathscr{L}$, in terms of wavelengths (its "electrical dimension") we write $\mathscr{L} = k\lambda$ and determine the length in wavelengths as

$$k = \frac{\mathscr{L}}{\lambda} = \frac{\mathscr{L}}{v}f$$

where we have substituted the wavelength in terms of the frequency and velocity of propagation as $\lambda = v/f$. Hence we obtain an important relation for the electrical length in terms of frequency and time delay:

$$\boxed{\begin{aligned} \frac{\mathscr{L}}{\lambda} &= f\frac{\mathscr{L}}{v} \\ &= fT_D \end{aligned}} \tag{1.8}$$

so that a dimension is one wavelength, $\mathscr{L}/\lambda = 1$, at a frequency that is the inverse of the time delay:

$$\boxed{f|_{\mathscr{L}=1\lambda} = \frac{1}{T_D}} \tag{1.9}$$

A *single-frequency sinusoidal* wave shifts phase as it travels a distance $\mathscr{L}$ of

$$
\begin{aligned}
\phi &= \beta\mathscr{L} \\
&= 2\pi\frac{\mathscr{L}}{\lambda} \qquad \text{rad} \\
&= \frac{\mathscr{L}}{\lambda} \times 360° \qquad \text{deg}
\end{aligned}
\tag{1.10}
$$

Hence if a wave travels a distance of one wavelength, $\mathscr{L} = 1\lambda$, it shifts phase by $\phi = 360°$. If the wave travels a distance of one-half wavelength, $\mathscr{L} = \frac{1}{2}\lambda$, it shifts phase by $\phi = 180°$. This can provide for cancellation, for example, when two antennas that are separated by a distance of one-half wavelength transmit the same frequency signal. Along a line containing the two antennas, the two radiating waves being of opposite phase cancel each other, giving a result of zero. This is essentially the reason that antennas have "patterns" where a null is produced in one direction, whereas a maximum is produced in another direction. Using this principle, phased-array radars "steer" their beams electronically rather than by rotating the antennas mechanically. Next consider a wave that travels a distance of one-tenth of a wavelength, $\mathscr{L} = \frac{1}{10}\lambda$. The phase shift incurred in doing so is only $\phi = 36°$, and a wave that travels one-one hundredth of a wavelength, $\mathscr{L} = \frac{1}{100}\lambda$, incurs a phase shift of $\phi = 3.6°$. Hence we say that

> *for any distance less than, say, $\mathscr{L} < \frac{1}{10}\lambda$, the phase shift is said to be negligible and the distance is said to be* <u>electrically</u> short.

For electric circuits whose physical dimension is electrically short, $\mathscr{L} < \frac{1}{10}\lambda$, Kirchhoff's voltage and current laws and other lumped-circuit analysis solution methods work very well.

> For physical dimensions that are NOT electrically short, Kirchhoff's laws and lumped-circuit analysis methods *give erroneous answers!*

For example, consider an electric circuit that is driven by a 10-kHz sinusoidal source. The wavelength at 10 kHz is 30 km (18.641 mi)! Hence at this frequency any circuit having a dimension of less than 3 km (1.86 mi) can be analyzed successfully using Kirchhoff's laws and lumped-circuit analysis methods. Electric power distribution systems operating at 60 Hz can be analyzed using Kirchhoff's laws and lumped-circuit analysis principles as long as their physical dimensions, such as transmission-line length, are less than some 310 mi! Similarly, a circuit driven by a 1-MHz sinusoidal source can be analyzed successfully using lumped-circuit analysis methods if its physical dimensions are less than 30 m! On the other hand, connection conductors in cell phone electronic circuits operating at a frequency of around 2 GHz cannot be analyzed using lumped-circuit analysis methods unless their dimensions are less than around 1.5 cm or about 0.6 in! We can, alternatively, determine the frequency where a dimension is electrically short in terms of the time delay from (1.8):

$$f|_{\mathscr{L}=(1/10)\lambda} = \frac{1}{10T_D} \tag{1.11}$$

Substituting $\lambda f = v$ into the time-delay expression in (1.4) gives the time delay as a portion of the period of the sinusoid, T:

$$
\begin{aligned}
T_D &= \frac{\mathscr{L}}{v} \\
&= \frac{\mathscr{L}}{\lambda}\frac{1}{f} \\
&= \frac{\mathscr{L}}{\lambda}T
\end{aligned}
\tag{1.12}
$$

where the period of the sinusoidal wave is $T = 1/f$. This shows that if we plot the current waves in Fig. 1.2 that enter and leave the connection leads versus time t *on the same time plot*, they will be displaced in time by a fraction of the period, $\mathscr{L}/\lambda$. If the length of the connection leads $\mathscr{L}$ is electrically short at this frequency, the two current waves will be displaced from each other *in time* by an inconsequential amount of less than $T/10$ and may be considered to be coincident in time. This is the reason that Kirchhoff's laws and lumped-circuit analysis methods work well only for circuits whose physical dimensions are "electrically small."

Waves propagated along transmission lines and radiated from antennas are of the same mathematical form as the currents on the connection leads of an element shown in (1.2). These are said to be *plane waves* where the electric and magnetic field vectors lie in a plane *transverse* or perpendicular to the direction of propagation of the wave, as shown in Fig. 1.4. These are said to be *transverse electromagnetic* (TEM) *waves*.

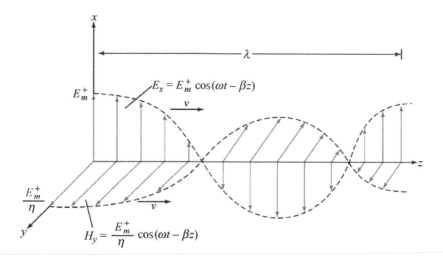

FIGURE 1.4. Electric and magnetic fields of plane waves on transmission lines and radiated by antennas.

This has demonstrated the following important principle in electromagnetics:

In electromagnetics, "physical dimensions" of structures don't matter; their "electrical dimensions in wavelengths" are important.

1.2   SPECTRAL (FREQUENCY) CONTENT OF DIGITAL WAVEFORMS AND THEIR BANDWIDTHS

A periodic waveform of fundamental frequency f_0 such as the digital clock waveform in Fig. 1.1 can be represented equivalently as an infinite summation of harmonically related sinusoids with the Fourier series shown in (1.1). The coefficients in the Fourier series are obtained for a digital clock waveform shown in Fig. 1.1, where the rise and fall times, τ_r and τ_f, are equal: $\tau_r = \tau_f$ (which digital clock waveforms approximate) as:

$$
\begin{aligned}
c_0 &= A\frac{\tau}{T} \\
c_n \angle \theta_n &= 2A\frac{\tau}{T}\frac{\sin(n\pi\tau/T)}{n\pi\tau/T}\frac{\sin(n\pi\tau_r/T)}{n\pi\tau_r/T} \angle -n\pi\frac{\tau+\tau_r}{T} \qquad \tau_r = \tau_f
\end{aligned}
$$

$$(1.13)$$

This result is in the form of the product of two $\sin(x)/x$ expressions, with the first depending on the ratio of the pulse width to the period, τ/T (also called the *duty cycle* of the waveform, $D = \tau/T$), and the second depending on the ratio of the pulse rise or fall time to the period, τ_r/T. [The *magnitude* of the coefficient, denoted as c_n, must be a positive number. Hence there may be an additional $\pm 180°$ added to the angle shown in (1.13), depending on the signs of each $\sin(x)$ term.] If, in addition to the rise and fall times being equal, the duty cycle is 50%, that is, the pulse is "on" for half the period and "off" for the other half of the period (which digital waveforms also tend to approximate), $\tau = \frac{1}{2}T$, the result for the coefficients given in (1.13) simplifies to

$$c_0 = \frac{A}{2}$$

$$c_n \angle \theta_n = A\frac{\sin(n\pi/2)}{n\pi/2}\frac{\sin(n\pi\tau_r/T)}{n\pi\tau_r/T} \angle -n\pi\left(\frac{1}{2}+\frac{\tau_r}{T}\right) \qquad \tau_r = \tau_f, \tau = \frac{T}{2}$$

Note that the first $\sin(x)/x$ function is zero for n even, so that *for equal rise and fall times and a 50% duty cycle the even harmonics are zero and the spectrum consists of only odd harmonics.* By replacing n/T with the smooth frequency variable f, $\boxed{n/T \to f}$, we obtain the *envelope* of the magnitudes of these discrete frequencies:

$$\boxed{c_n = 2A\frac{\tau}{T}\left|\frac{\sin(\pi f\tau)}{\pi f\tau}\right|\left|\frac{\sin(\pi f\tau_r)}{\pi f\tau_r}\right| \qquad \tau_r = \tau_f, \frac{n}{T} \to f} \qquad (1.14)$$

In doing so, remember that the spectral components occur only at the discrete frequencies $f_0, 2f_0, 3f_0, \ldots$.

Observe some important properties of the $\sin(x)/x$ function:

$$\lim_{x \to 0} \frac{\sin(x)}{x} = 1$$

which relies on the property that $\sin(x) \cong x$ for small x (or using l'Hôpital's rule) and

$$\left|\frac{\sin(x)}{x}\right| \leq \begin{cases} 1 & x \leq 1 \\ \frac{1}{x} & x \geq 1 \end{cases}$$

The second property allows us to obtain a *bound* on the magnitudes of the c_n coefficients and relies on the fact that $|\sin(x)| \leq 1$ for all x.

A *square wave* is the trapezoidal waveform where the rise and fall times are *zero*:

$$c_0 = A\frac{\tau}{T}$$
$$c_n \angle \theta_n = 2A\frac{\tau}{T}\frac{\sin(n\pi\tau/T)}{n\pi\tau/T} \angle -n\pi\frac{\tau}{T} \qquad \tau_r = \tau_f = 0$$

If the duty cycle of the square wave is 50%, this result simplifies to

$$c_0 = \frac{A}{2}$$
$$c_n \angle \theta_n = \begin{cases} \dfrac{2A}{n\pi} \angle -\dfrac{\pi}{2} & n \text{ odd} \\ 0 & n \text{ even} \end{cases} \qquad \tau_r = \tau_f = 0, \; \tau = \frac{T}{2}$$

Figure 1.5 shows a plot of the magnitudes of the c_n coefficients for a square wave where the rise and fall times are zero, $\tau_r = \tau_f = 0$. The spectral components appear only at *discrete* frequencies, $f_0, 2f_0, 3f_0, \ldots$. The envelope is shown with a dashed line. Observe that the envelope goes to zero where the argument of $\sin(\pi f\tau)$ becomes a multiple of π at $f = 1/\tau, 2/\tau, \ldots$.

A more useful way of plotting the envelope of the magnitudes of the spectral coefficients is by plotting the horizontal frequency axis logarithmically and, similarly, plotting the magnitudes of the coefficients along the vertical axis in decibels as $|c_n|_{\text{dB}} = 20\log_{10}|c_n|$. The envelope as well as the bounds of the magnitudes of the $\sin(x)/x$ function are shown in Fig. 1.6.

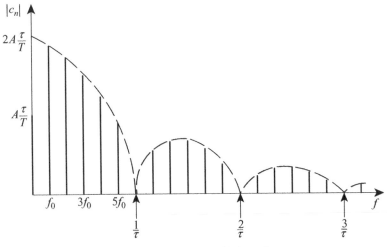

FIGURE 1.5. Plot of the magnitudes of the c_n coefficients for a square wave, $\tau_r = \tau_f = 0$.

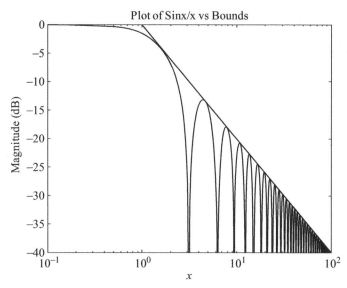

FIGURE 1.6. The envelope and bounds of the $\sin(x)/x$ function are plotted with logarithmic axes.

Observe that the actual result is bounded by 1 for $x \leq 1$ and decreases at a rate of -20 dB/decade for $x \geq 1$. This rate is equivalent to a $1/x$ decrease. Also note that the magnitudes of the actual spectral components go to zero where the argument of $\sin(x)$ goes to a multiple of π or $x = \pi, 2\pi, 3\pi, \ldots$.

The amplitude of the spectral components of a trapezoidal waveform where $\tau_r = \tau_f$ given in (1.14) is the product of two $\sin(x)/x$ functions: $\sin(x_1)/x_1 \times \sin(x_2)/x_2$. When log-log axes are used, this gives the result for the bounds on the amplitudes of the spectral coefficients shown in Fig. 1.7. Note that the bounds are constant (0 dB/decade) out to the first breakpoint of $f_1 = 1/\pi\tau = f_0/\pi D$, where the *duty cycle* is $D = \tau/T = \tau f_0$. Above this they decrease at a rate of -20 dB/decade out to a second breakpoint of $f_2 = 1/\pi\tau_r$ and decrease at a rate of -40 dB/decade above that. This plot shows the important result that *the high-frequency spectral content of the trapezoidal clock waveform is determined by the pulse rise and fall times.* Longer rise and fall times push the second breakpoint lower in frequency, thereby reducing the high-frequency spectral content. Shorter rise and fall times push the second breakpoint higher in frequency, thereby increasing the high-frequency spectral content.

How do we quantitatively determine the *bandwidth* of a periodic clock waveform? Although the Fourier series in (1.1) requires that we sum an *infinite* number of terms, as a practical matter we use NH terms (harmonics) as an approximate finite-term approximation: $\tilde{x}(t) = c_0 + \sum_{n=1}^{NH} c_n \cos(n\omega_0 t + \theta_n)$. The *pointwise approximation error* is $x(t) - \tilde{x}(t)$. The logical definition of

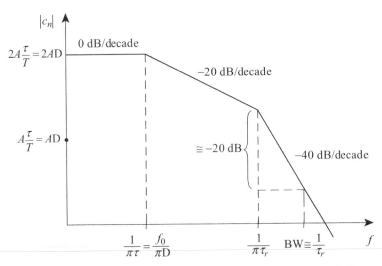

FIGURE 1.7. Bounds on the spectral coefficients of the trapezoidal pulse train for equal rise and fall times $\tau_r = \tau_f$.

the *bandwidth* (BW) of the waveform is that the BW should be the *significant spectral content of the waveform*. In other words,

> *the BW should be the minimum number of harmonic terms required to reconstruct the original periodic waveform such that adding more harmonics gives a negligible reduction in the pointwise error, whereas using less harmonics gives an excessive pointwise reconstruction error.*

If we look at the plot of the bounds on the magnitude spectrum shown in Fig. 1.7, we see that above the second breakpoint, $f_2 = 1/\pi\tau_r$, the levels of the harmonics are rolling off at a rate of -40 dB/decade. If we go past this second breakpoint by a factor of about 3 to a frequency that is the inverse of the rise and fall time, $f = 1/\tau_r$, the levels of the component at the second breakpoint will have been reduced further, by around 20 dB. Hence above this frequency the remaining frequency components are probably so small in magnitude that they do not provide any substantial contribution to the shape of the resulting waveform. Hence we might define the *bandwidth* of the trapezoidal clock waveform (and other data waveforms of similar shape) to be

$$BW \cong \frac{1}{\tau_r} \qquad\qquad (1.15)$$

TABLE 1.2. Spectral (Frequency) Components of a 5-V, 1-GHz, 50% Duty Cycle, 100-ps Rise/Fall Time Digital Clock Signal

Harmonic	Frequency (GHz)	Wavelength, λ (cm)	Level (V)	Angle (deg)
1	1	30	3.131	−108
3	3	10	0.9108	−144
5	5	6	0.4053	−180
7	7	4.29	0.1673	144
9	9	3.33	0.0387	108

The bandwidth in (1.15) obviously does not apply to a square wave, $\tau_r = \tau_f = 0$, since that would imply that its BW would be infinite. But an ideal square wave cannot be constructed in practice.

For a 1-GHz clock waveform having a 5-V amplitude, a 50% duty cycle, and 100-ps rise and fall times, the bandwidth by this criterion is 10 GHz. Table 1.2 shows the first nine coefficients for this digital waveform. Observe that the ninth harmonic of 9 GHz has a wavelength of 3.33 cm. Using Kirchhoff's voltage and current laws and lumped-circuit analysis principles to analyze a circuit driven by this frequency would require that the largest dimension of the circuit be less that 3.33 mm (0.131 in)! Similarly, to analyze a circuit that is driven by the fundamental frequency of 1 GHz whose wavelength is 30 cm using Kirchhoff's laws and lumped-circuit analysis methods would restrict the maximum circuit dimensions to being less than 3 cm or about 1 in (2.54 cm)! This shows that the use of lumped-circuit analysis methods to analyze a circuit having a physical dimension of, say, 1 in that is driven by this clock waveform would result in erroneous results for all but perhaps the fundamental frequency of the waveform! Figure 1.8 shows the bounds and envelope of the spectrum for this waveform. The first breakpoint of $f_1 = 1/\pi\tau = f_0/\pi D = 636.6$ MHz is not shown because it falls below the fundamental frequency of 1 GHz.

Figure 1.9(a) to (d) show the approximation to the clock waveform achieved by adding the dc component and the first three harmonics, the first five harmonics, the first seven harmonics, and the first nine harmonics, respectively. This increasing convergence of these partial sums to the true waveform supports the idea that using only the first 10 harmonic components as its BW, BW $= 1/\tau_r = 1/0.1$ ns $= 10$ GHz, gives a reasonable representation of the actual waveform.

This Fourier representation of a periodic waveform, such as a digital waveform, as a *summation* of single-frequency sinusoidal basic components as in (1.1) provides a useful and simple method for approximately solving a linear system indirectly. Consider the single-input, $x(t)$, single-output, $y(t)$,

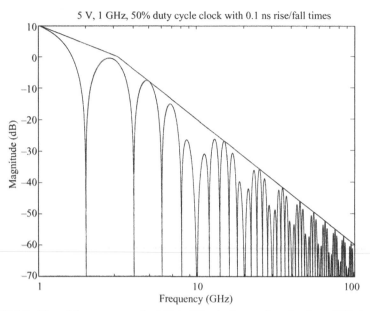

FIGURE 1.8. Plot of the spectrum of a 5-V, 1-GHz clock waveform having a 50% duty cycle and rise and fall times of 0.1 ns.

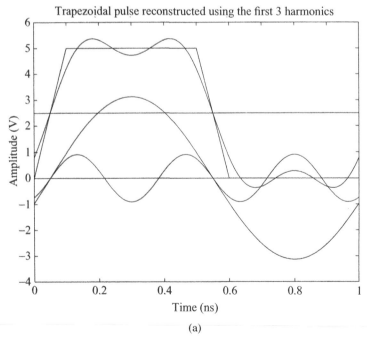

(a)

FIGURE 1.9. Approximating the clock waveform using (a) the first three harmonics, (b) the first five harmonics, (c) the first seven harmonics, and (d) the first nine harmonics.

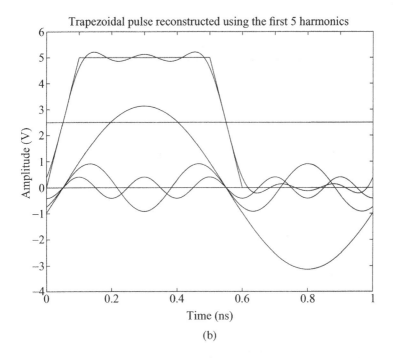

(b)

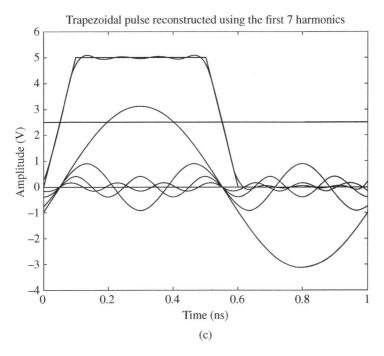

(c)

FIGURE 1.9. (*Continued*)

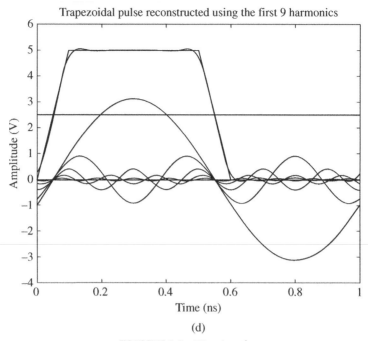

(d)

FIGURE 1.9. (*Continued*)

linear system illustrated in Fig. 1.10. A linear system is one for which the principle of *superposition* applies. In other words, the system is *linear* if $x_1(t) \rightarrow y_1(t)$ and $x_2(t) \rightarrow y_2(t)$, then (1) $x_1(t) + x_2(t) \rightarrow y_1(t) + y_2(t)$ and (2) $kx(t) \rightarrow ky(t)$. The output is related to the input with a differential equation:

$$\frac{d^n y(t)}{dt} + a_1 \frac{d^{(n-1)} y(t)}{dt} + \cdots + a_n y(t) = b_0 \frac{d^m x(t)}{dt}$$

$$+ b_1 \frac{d^{(m-1)} x(t)}{dt} + \cdots + b_m x(t)$$

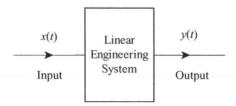

FIGURE 1.10. Single-input, single-output linear system.

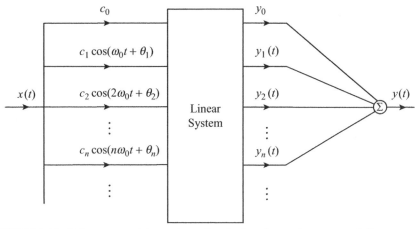

FIGURE 1.11. Using superposition to determine the (steady-state) response of a linear system to a waveform by passing the individual Fourier components through the system and summing their responses at the output.

The differential equation relating the input and output (sometimes referred to as the *transfer function*) can be solved for the waveform of the output, $y(t)$. But this can be a difficult and tedious task.

A simpler but approximate solution method is represented in Fig. 1.11. Decompose the input waveform, $x(t)$, into its Fourier components and pass each one through the system, giving a response to that component. Then sum all these responses in *time*, which gives an approximate solution to $y(t)$. This is, for several reasons a much simpler solution process than the direct solution of the differential equation relating the input and output. The basic functions in the Fourier series are the sinusoids: $c_n \cos(n\omega_0 t + \theta_n)$. It is usually much easier to determine the response to each of these sinusoids (referred to as the *frequency domain*). Then these responses are *summed in time* to give an approximation to the output, $y(t)$. An important restriction to this method is that it neglects any transient part of the solution and gives only the steady-state response.

As an example of this powerful technique, consider an RC circuit that is driven by a periodic square-wave voltage source as shown in Fig. 1.12. The square wave has an amplitude of 1 V, a period of 2 s, and a pulse width of 1 s (50% duty cycle). The RC circuit, which consists of the series connection of $R = 1\,\Omega$ and $C = 1\,\text{F}$ has a time constant of $RC = 1\text{s}$, and the voltage across the capacitor is the desired output voltage of this linear "system." The nodes of the circuit are numbered in preparation for using the SPICE circuit analysis program (or the personal computer version, PSPICE) to analyze it and plot the exact solution. The Fourier series of the input, $V_S(t)$, using only the first seven harmonics, is ($\omega_0 = 2\pi/T = \pi$)

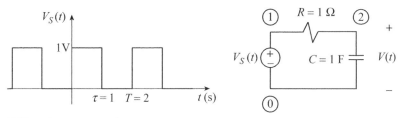

FIGURE 1.12. Example of using superposition of the Fourier components of a signal in obtaining the (steady-state) response to that signal.

$$V_S(t) = c_0 + c_1 \cos(\omega_0 t + \theta_1) + c_3 \cos(3\omega_0 t + \theta_3) + c_5 \cos(5\omega_0 t + \theta_5)$$

$$+ c_7 \cos(7\omega_0 t + \theta_7)$$

$$= \frac{1}{2} + \frac{2}{\pi} \cos(\pi t - 90°) + \frac{2}{3\pi} \cos(3\pi t - 90°) + \frac{2}{5\pi} \cos(5\pi t - 90°)$$

$$+ \frac{2}{7\pi} \cos(7\pi t - 90°)$$

To determine the Fourier series of the output we first determine the response to a single-frequency input, $x(t) = c_n \cos(n\omega_0 t + \theta_n)$. The ratio of the output and this single-frequency sinusoidal input is referred to as the *transfer function* of the linear system response to this single-frequency input. The phasor (sinusoidal steady-state) *transfer function* of this linear system is the ratio of the output and input (magnitude and phase):

$$\hat{H}(jn\omega_0) = \frac{\hat{V}}{\hat{V}_S}$$

$$= \frac{1}{1 + jn\omega_0 RC}$$

$$= \frac{1}{1 + jn\pi}$$

$$= \frac{1}{\sqrt{1 + (n\pi)^2}} \angle - \tan^{-1}(n\pi)$$

$$= H_n \angle \phi_n$$

The phasor (sinusoidal steady state) voltages and currents will be denoted with carets and are complex valued, having a magnitude and an angle: $\hat{V} = V \angle \theta_V$ and $\hat{I} = I \angle \theta_I$. The output of this "linear system" is the voltage across the capacitor, $V(t)$, whose Fourier coefficients are obtained as $c_n H_n \angle (\theta_n + \phi_n)$

giving the Fourier series of the time-domain output waveform as

$$V(t) = c_0 H_0 + \sum_{n=1}^{7} c_n H_n \cos[n\omega_0 t + \angle(\theta_n + \phi_n)]$$

$$= 0.5 + 0.1931 \cos(\pi t - 162.34°) + 0.0224 \cos(3\pi t - 173.94°)$$

$$+ 0.0081 \cos(5\pi t - 176.36°) + 0.0041 \cos(7\pi t - 177.4°)$$

Figure 1.13 shows the approximation to the output waveform for $V(t)$ obtained by summing in time the steady-state responses to only the dc component and the first seven harmonics of $V_S(t)$.

The exact result for $V(t)$ is obtained with PSPICE and shown in Fig. 1.14. The PSPICE program is

```
EXAMPLE
VS 1 0 PULSE(0 1 0 0 0 1 2)
RS 1 2 1
C 2 0 1
.TRAN 0.01 10 0 0.01
.PRINT TRAN V(2)
.PROBE
.END
```

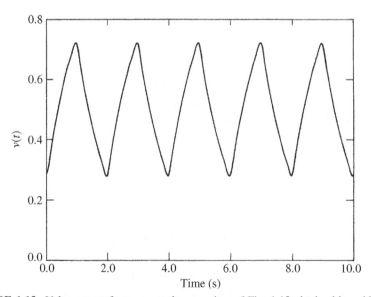

FIGURE 1.13. Voltage waveform across the capacitor of Fig. 1.12 obtained by adding the (steady-state) responses of the dc component and the first seven harmonics of the Fourier series of the square wave.

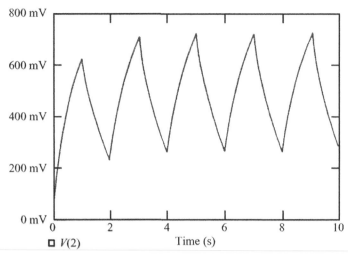

FIGURE 1.14. PSPICE solution for $V(t)$ for the circuit in Fig. 1.12.

Note that there is an initial transient part of the solution over the first 2 or 3 s due to the capacitor being charged up to its steady-state voltage. These results make sense because as the square wave transitions to 1 V, the voltage across the capacitor increases according to $1 - e^{-t/RC}$. Since the time constant is $RC = 1$ s, the voltage has not reached steady state (which requires about five time constants to have elapsed) when the square wave turns off at $t = 1$ s. Then the capacitor voltage begins to discharge. But when the square wave turns on again at $t = 2$ s, the capacitor has not fully discharged and begins recharging. This process and the resulting output voltage waveform repeats with a period of 2 s. The transitions in the exact waveform of the output voltage in Fig. 1.14 are sharper than the corresponding transitions in the approximate waveform in Fig. 1.13 obtained by summing the responses to the first seven harmonics of the Fourier series of the input waveform. This is due to neglecting the responses to the high-frequency components of the input waveform and is a general property. The initial transient response in the exact PSPICE solution in Fig. 1.14 is absent from the Fourier method in Fig. 1.13 since the Fourier method only obtains the steady-state response.

1.3   THE BASIC TRANSMISSION-LINE PROBLEM

The basic transmission-line problem connects a *source* to a *load* with a *transmission line* as shown in Fig. 1.15(a). The transmission line consists of a parallel pair of conductors of total length $\mathscr{L}$ having uniform cross sections along its length. The objective will be to determine the time-

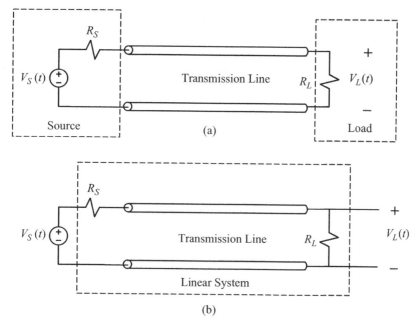

FIGURE 1.15. Basic transmission-line problem.

domain response waveform of the output voltage of the line, $V_L(t)$, given the termination impedances, R_S and R_L, the source voltage waveform, $V_S(t)$, and the properties of the transmission line. If the source and termination impedances are linear, we may alternatively view the transmission-line problem as a linear system having an input $V_S(t)$ and an output $V_L(t)$ by embedding the terminations and the transmission line into one system, as shown in Fig. 1.15(b).

We first determine the *frequency-domain* response of the system as shown in Fig. 1.16. A single-frequency sinusoidal source, $V_S(t) = V_S \cos(\omega t + \theta_S)$, produces a similar form of a sinusoidal load voltage: $V_L(t) = V_L \cos(\omega t + \theta_L)$.

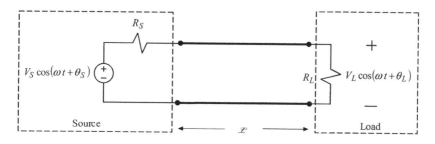

FIGURE 1.16. General source–load configuration.

The source and load are separated by a parallel pair of wires or a pair of lands of length $\mathscr{L}$. The lumped-circuit model *ignores* the two interconnect conductors of length $\mathscr{L}$. Analyzing this configuration as a lumped circuit gives (using voltage division and ignoring the interconnect conductors) the ratio of the source and load voltage magnitudes as

$$\frac{V_L}{V_S} = \frac{R_L}{R_S + R_L}$$

and the phase angles are identical: $\theta_S = \theta_L$. These, according to a lumped-circuit model of the line, remain the same for *all source frequencies*!

Consider the specific configuration shown in Fig. 1.17. The parameters are $R_S = 10\,\Omega$ and $R_L = 1000\,\Omega$ for a line of total length of $\mathscr{L} = 0.3$ m (or about 12 in). Ignoring the effects of the interconnect conductors gives $V_L/V_S = 0.99$, and the phases are related as $\theta_L - \theta_S = 0°$. The *exact solution* is obtained by including the two interconnect conductors of length $\mathscr{L}$ as a distributed-parameter *transmission line*. The circuit analysis computer program, PSPICE, contains an exact transmission-line model of the interconnect conductors. Figure 1.18 shows the *exact* ratio of the voltage magnitude, V_L/V_S, and voltage angle, $\theta_L - \theta_S$, versus the frequency of the source as it is swept in frequency from 1 MHz to 1 GHz. Model the interconnect conductors as a distributed-parameter transmission line having a characteristic impedance of $Z_C = 50\,\Omega$ and a one-way delay of the interconnect line of

$$T_D = \frac{\mathscr{L} = 0.3\text{ m}}{v_0 = 3 \times 10^8 \text{ m/s}} = 1 \text{ ns}$$

The entire configuration is analyzed using PSPICE. The AC mode of analysis in PSPICE is used to obtain the *frequency-domain* transfer function of this system. The PSPICE program is

```
EXAMPLE
VS 1 0 AC 1 0
RS 1 2 10
T 2 0 3 0 Z0=50 TD=1N
RL 3 0 1K
.AC DEC 50 1MEG 1G
.PRINT AC VM(1) VP(1) VM(3) VP(3)
.PROBE
.END
```

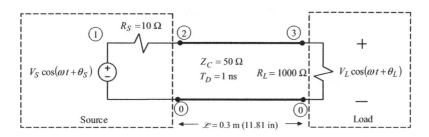

FIGURE 1.17. Specific example treating the connection lands as a transmission line.

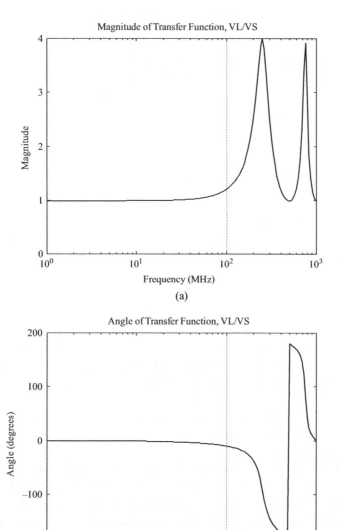

(a)

(b)

FIGURE 1.18. Frequency response of the line in Fig. 1.17.

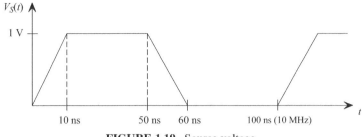

FIGURE 1.19. Source voltage.

Figure 1.18 shows that the magnitudes and angles of the *transfer function* voltages, V_L/V_S and $\theta_L - \theta_S$, begin to deviate rather drastically from the low-frequency lumped-circuit analysis result of $V_L/V_S = 0.99$ and $\theta_L - \theta_S = 0°$ above about 100 MHz. The line is one-tenth of a wavelength (electrically short) at

$$f|_{\mathscr{L}=(1/10)\lambda} = \frac{1}{10T_D = 10\,\text{ns}} = 100\,\text{MHz}$$

(denoted by the vertical line at 100 MHz in both plots). This is evident in the plots in Fig. 1.18. Hence the interconnect line is electrically long above 100 MHz. The interconnect line is one wavelength at 1 GHz:

$$f|_{\mathscr{L}=\lambda} = \frac{1}{T_D = 1\,\text{ns}} = 1\,\text{GHz}$$

Observe that the magnitude plot in Fig. 1.18(a) shows two peaks of 250 MHz and 750 MHz where the interconnect line electrical length is $\lambda/4$ and $\frac{3}{4}\lambda$, respectively, and the magnitude of the transfer function increases to a level of 4. There are two minima at 500 MHz and 1 GHz, where the interconnect line electrical length is $\lambda/2$ and λ, respectively. Above 1 GHz (the last frequency plotted) the pattern replicates, which is a general property of transmission lines.

Finally, we investigate the *time-domain* response of the line where we drive the line with a clock signal of 10 MHz fundamental frequency (a period of 100 ns), an amplitude of 1 V, rise and fall times of 10 ns, and a 50% duty cycle as shown in Fig. 1.19. It is typical for the rise and fall times of digital waveforms to be chosen to be around 10% of the period T in order to give adequate "setup" and "hold" times. The exact *time-domain* load voltage waveform, $V_L(t)$, is obtained with the .TRAN module of PSPICE for the waveform in Fig. 1.19. The PSPICE program used is

```
EXAMPLE
VS 1 0 PULSE(0 1 0 10N 10N 40N 100N)
RS 1 2 10
T 2 0 3 0 Z0=50 TD=1N
RL 3 0 1K
.TRAN 0.1N 200N 0 0.1N
.PRINT TRAN V(1) V(3)
.PROBE
.END
```

Figure 1.20 shows a comparison of the load voltage waveform, $V_L(t)$, and the source voltage waveform, $V_S(t)$, for this source waveform over two cycles of the source. The source voltage and load voltage waveforms are virtually identical, and the interconnect line clearly has no substantial effect. From the frequency response of the waveform in Fig. 1.18, we see that the first 10 harmonics of this waveform (the bandwidth of the waveform is $BW = 1/\tau_r = 100\,MHz$)—10, 20, 30, 40, 50, 60, 70, 80, 90, 100 MHz—all fall below the frequency where the line ceases to be electrically short: 100 MHz. This is what we expect when the major harmonic components of the waveform (its BW) fall into the frequency range where the line is electrically short for all of them.

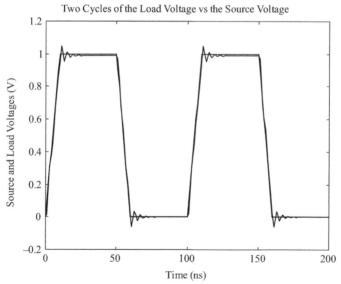

FIGURE 1.20. Comparison of the source and load waveforms for a 1-V, 10-MHz waveform with rise and fall times of $\tau_r = \tau_f = 10\,ns$ and a 50% duty cycle (see Fig. 1.19).

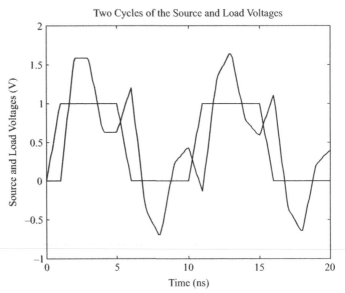

FIGURE 1.21. Comparison of the source and load waveforms for a 1-V, 100-MHz waveform with rise and fall times of $\tau_r = \tau_f = 1$ ns and a 50% duty cycle.

Figure 1.21 shows the same comparison when the source parameters are changed to a 100-MHz fundamental frequency (a period of 10 ns) having an amplitude of 1 V, rise and fall times of 1 ns, and a 50% duty cycle. The PSPICE program is changed slightly just by changing the parameters of the "PULSE" function and the ".TRAN" lines to

```
EXAMPLE
VS 1 0 PULSE(0 1 0 1N 1N 4N 10N)
RS 1 2 10
T 2 0 3 0 Z0=50 TD=1N
RL 3 0 1K
.TRAN 0.01N 20N 0 0.01N
.PRINT TRAN V(1) V(3)
.PROBE
.END
```

From Fig. 1.18, this waveform contains the first 10 harmonics that constitute the major components in its bandwidth (BW $= 1/\tau_r = 1$ GHz): 100 MHz, 200 MHz, 300 MHz, 400 MHz, 500 MHz, 600 MHz, 700 MHz, 800 MHz, 900 MHz, and 1 GHz. The line length is $\lambda/10$ at its fundamental frequency, 100 MHz, and 1λ at its tenth harmonic of 1 GHz. Observe that the load voltage waveform bears no resemblance to the source waveform. From the frequency response of the system in Figure 1.18, we see that all of these harmonics fall in the frequency range where the interconnect line is *electrically long* (>100 MHz), so this is expected.

This has shown that as the frequencies of the sources increase to the point where the interconnect lines connecting the source and the load become *electrically long*, the standard lumped-circuit models are no longer valid and give erroneous answers. The requirement to model electrically long interconnects requires that we master transmission-line modeling.

2

TIME-DOMAIN ANALYSIS OF TWO-CONDUCTOR LINES

In this chapter we study the time-domain solution of transmission lines composed of two conductors of length $\mathscr{L}$, as shown in Fig. 2.1. The two conductors are assumed to be parallel to the z axis of a rectangular coordinate system and of uniform cross section along their lengths. The source driving the line, $V_S(t)$, may have some general waveshape in time, although we will generally concentrate on periodic waveforms of trapezoidal shape as in Fig. 1.1, representing digital clock and data signals. The source will be represented as a Thévenin equivalent circuit consisting of a voltage source $V_S(t)$ in series with a source resistor represented by R_S. The load will be represented by a resistance R_L, although PSPICE can be used to investigate lines terminated at either end in nonlinear dynamic terminations. The voltage and current along the line will be a function of position along the line, z, and time, t, $V(z, t)$ and $I(z, t)$. Given the parameters of the source waveform and the values of the source and load terminations along with the dimensions and properties of the line, we will be interested in determining the solutions for the voltage and current waveforms at the input and output of the line: $V(0, t)$, $V(\mathscr{L}, t)$, $I(0, t)$, and $I(\mathscr{L}, t)$.

Transmission Lines in Digital Systems for EMC Practitioners, First Edition.
Clayton R. Paul.
© 2012 John Wiley & Sons, Inc. Published 2012 by John Wiley & Sons, Inc.

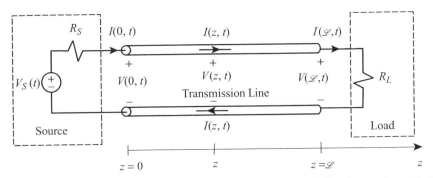

FIGURE 2.1. General configuration of the two-conductor transmission lines to be studied.

2.1 THE TRANSVERSE ELECTROMAGNETIC MODE OF PROPAGATION AND THE TRANSMISSION-LINE EQUATIONS

We consider *lossless lines* in which the two conductors and the surrounding medium are considered to be lossless. The two conductors are parallel to each other and the z axis and have general cross sections that are uniform along their z-length axis as shown in Fig. 2.2. They are said to be *uniform lines*.

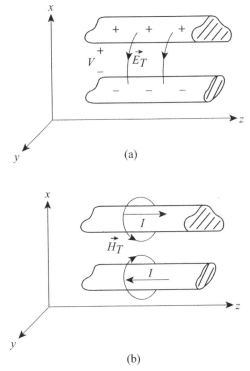

FIGURE 2.2. Uniform transmission lines of general cross section.

If we apply a voltage between the two conductors, an electric field, $\vec{\mathbf{E}}_T$, will be developed between them that lies solely in the x–y plane. If we pass a current down one conductor and return on the other conductor, a magnetic field, $\vec{\mathbf{H}}_T$, will developed (by the right-hand rule) that lies solely in the x–y plane and passes through the loop between them. This is said to be the transverse electromagnetic (TEM) mode of propagation that propagates down the line from the source to the load. This electromagnetic field propagates the source signal, $V_S(t)$, from the source to the load. These are said to be *plane waves* since the electromagnetic fields lie in the x–y plane *transverse* to the direction of propagation.

The voltage places a plus charge on the upper conductor and an equal but negative charge on the lower conductor and hence separates charge. Hence the two conductors separate charge and therefore represent a capacitance. We assume that the two conductors are infinitely long so that we can neglect fringing of the fields at the two ends of the conductors. Similarly, the current passing down the upper conductor and returning on the lower conductor causes a magnetic field to penetrate the loop between the two conductors and hence represents an inductance. Since the two conductors are assumed to be infinitely long, we can represent the two conductors with a *per-unit-length* capacitance c F/m and a *per-unit-length* inductance l H/m. Since the line is uniform, we can divide the line into Δz sections that are *electrically small* and represent it with a lumped equivalent circuit as shown in Fig. 2.3. Observe that we represent the total inductance and capacitance over each Δz segment with $c\,\Delta z$ and $l\,\Delta z$.

For this lumped equivalent circuit to be valid for all frequencies of the source, we let $\Delta z \to 0$ and obtain the *transmission-line equations* from Fig. 2.4. First write KVL as

$$V(z + \Delta z, t) - V(z, t) = -l\,\Delta z \frac{\partial I(z, t)}{\partial t} \tag{2.1a}$$

and write KCL as

$$I(z + \Delta z, t) - I(z, t) = -c\,\Delta z \frac{\partial V(z + \Delta z, t)}{\partial t} \tag{2.1b}$$

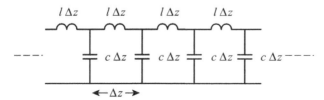

FIGURE 2.3. Representing the line with electrically short segments.

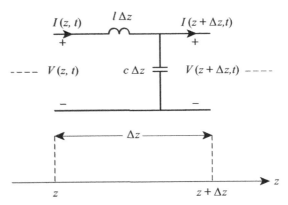

FIGURE 2.4. Per-unit-unit-length equivalent circuit.

Dividing both sides by Δz gives

$$\frac{V(z + \Delta z, t) - V(z, t)}{\Delta z} = -l\frac{\partial I(z, t)}{\partial t} \qquad (2.2\text{a})$$

$$\frac{I(z + \Delta z, t) - I(z, t)}{\Delta z} = -c\frac{\partial V(z + \Delta z, t)}{\partial t} \qquad (2.2\text{b})$$

and letting $\Delta z \to 0$, we have

$$\lim_{\Delta z \to 0} \frac{V(z + \Delta z, t) - V(z, t)}{\Delta z} = \frac{\partial V(z, t)}{\partial z} \qquad (2.3\text{a})$$

$$\lim_{\Delta z \to 0} \frac{I(z + \Delta z, t) - I(z, t)}{\Delta z} = \frac{\partial I(z, t)}{\partial z} \qquad (2.3\text{b})$$

we obtain the *transmission-line equations*:

$$\boxed{\frac{\partial V(z, t)}{\partial z} = -l\frac{\partial I(z, t)}{\partial t}} \qquad (2.4\text{a})$$

$$\boxed{\frac{\partial I(z, t)}{\partial z} = -c\frac{\partial V(z, t)}{\partial t}} \qquad (2.4\text{b})$$

which comprise a set of coupled partial differential equations. We can "uncouple" these coupled partial differential equations by differentiating one with respect to z and the other with respect to t:

$$\frac{\partial^2 V(z, t)}{\partial z^2} = -l\frac{\partial^2 I(z, t)}{\partial z\, \partial t} \tag{2.5a}$$

$$\frac{\partial^2 I(z, t)}{\partial t\, \partial z} = -c\frac{\partial^2 V(z, t)}{\partial t^2} \tag{2.5b}$$

and substitute to obtain the "uncoupled" second-order differential equations which we are to solve:

$$\boxed{\frac{\partial^2 V(z, t)}{\partial z^2} = \underbrace{lc}_{1/v^2}\frac{\partial^2 V(z, t)}{\partial t^2}} \tag{2.6a}$$

$$\boxed{\frac{\partial^2 I(z, t)}{\partial z^2} = \underbrace{cl}_{1/v^2}\frac{\partial^2 I(z, t)}{\partial t^2}} \tag{2.6b}$$

The velocity of propagation of the TEM waves along the line is

$$\boxed{v = \frac{1}{\sqrt{lc}} \qquad \frac{\text{m}}{\text{s}}} \tag{2.7}$$

So we can obtain one parameter from the other:

$$\boxed{l = \frac{1}{cv^2}} \tag{2.8a}$$

$$\boxed{c = \frac{1}{lv^2}} \tag{2.8b}$$

If the medium surrounding the two conductors is *homogeneous* with uniform permittivity and permeability ε and μ, respectively, we have the identity

$$lc = \mu\varepsilon \qquad \text{homogeneous medium} \qquad (2.9)$$

in which case

$$v = \frac{1}{\sqrt{\mu\varepsilon}} \qquad \text{homogeneous medium} \qquad (2.10)$$

The general solution to the transmission-line equations is

$$V(z,t) = V^{+}\left(t - \frac{z}{v}\right) + V^{-}\left(t + \frac{z}{v}\right) \qquad (2.11a)$$

$$I(z,t) = \frac{1}{Z_C}V^{+}\left(t - \frac{z}{v}\right) - \frac{1}{Z_C}V^{-}\left(t + \frac{z}{v}\right) \qquad (2.11a)$$

where Z_C is the *characteristic impedance of the line*:

$$\begin{aligned} Z_C &= \sqrt{\frac{l}{c}} \quad \Omega \\ &= vl \\ &= \frac{1}{vc} \end{aligned} \qquad (2.12)$$

The V^{+} and V^{-} are, as yet, undetermined *functions* but depend on z, t, and v *only* as $t + z/v$ and $t - z/v$. These functions are determined by the source and load $V_S(t)$, R_S, and R_L. Also note that there is an important negative sign in the solution for the current. The V^{+} represent *forward-traveling waves* traveling in the $+z$ direction, whereas the V^{-} represent *backward-traveling waves* traveling in the $-z$ direction. So, in general, we have waves of voltage and current (or, equivalently, waves of electric and magnetic fields) traveling back and forth down the line. We see that the voltage and current waves are in general being reflected at the source and at the load, and the combination of these waves determines the total voltage and current at the source and the load ends of the line.

2.2   THE PER-UNIT-LENGTH PARAMETERS

The transmission-line equations are identical in form for all transmission lines. What distinguishes one line from another? All line cross-sectional dimensions and properties are contained in the per-unit-length parameters l and c. So we must first obtain equations for l and c for each specific line cross section. We consider two general types of line cross sections: wire-type lines consisting of conductors of circular cylindrical cross section (wires) and lines whose conductors have rectangular cross sections (lands), which are typical on printed circuit boards (PCBs).

2.2.1   Wire-Type Lines

Wire-type lines consist of conductors of circular cylindrical cross section commonly called "wires." These common configurations are illustrated in Fig. 2.5. Figure 2.5(a) shows a two-wire line, Fig. 2.5(b) shows one wire above an infinite perfectly conducting ground plane, and Fig. 2.5(c) shows a coaxial cable.

First we determine the dc magnetic field of an infinitely long isolated wire of radius r_w, as shown in Fig. 2.6. The dc magnetic field of an isolated wire both inside and outside the wire can be determined using Ampère's law:

$$\oint_c \vec{\mathbf{H}} \cdot d\vec{\mathbf{l}} = I_{\text{enclosed}} \qquad (2.13)$$

Utilizing symmetry we construct a circular contour c of radius r inside the wire and obtain

$$
\begin{aligned}
\oint_c \vec{\mathbf{H}} \cdot d\vec{\mathbf{l}} &= H_\phi \, 2\pi r \\
&= I_{\text{enclosed}} \\
&= \frac{I}{\pi r_w^2} \pi r^2 \\
&= I \frac{r^2}{r_w^2}
\end{aligned}
\qquad (2.14)
$$

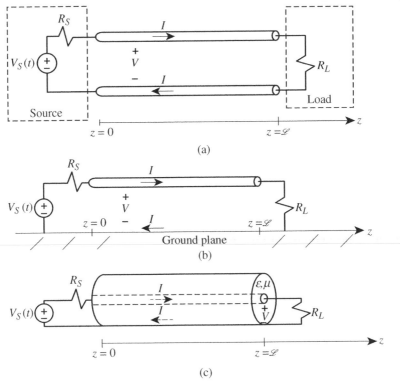

FIGURE 2.5. Common wire-type lines.

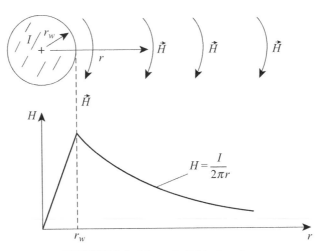

FIGURE 2.6. Magnetic field of a wire.

so that

$$B_\phi = \mu_0 H_\phi$$

$$= \frac{\mu_0 I r}{2\pi r_w^2} \quad \frac{\text{Wb}}{\text{m}^2} \qquad r < r_w \qquad (2.15)$$

and $\mu_0 = 4\pi \times 10^{-7}\text{H/m}$ is the permeability of free space. Constructing a circular contour of radius r outside the wire and utilizing symmetry gives

$$\oint_c \vec{H} \cdot d\vec{l} = H_\phi\, 2\pi r$$

$$= I_{\text{enclosed}} \qquad (2.16)$$

$$= I$$

so that

$$B_\phi = \mu_0 H_\phi$$

$$= \frac{\mu_0 I}{2\pi r} \quad \frac{\text{Wb}}{\text{m}^2} \qquad r > r_w \qquad (2.17)$$

These are plotted in Fig. 2.6.

Next we determine the magnetic flux through a planar surface of length $\mathscr{L}$ that is parallel to the wire and has edges of radii $R_2 > R_1$ from the the wire center axis. This is shown in Fig. 2.7. Gauss's law provides that the magnetic flux through a closed surface is zero; that is, all magnetic field lines must form closed loops:

$$\oint_s \vec{B} \cdot d\vec{s} = 0 \qquad (2.18)$$

Form the closed wedge-shaped surface shown in Fig. 2.7(b). Since the magnetic field is tangent to the ends of this wedge-shaped closed surface and is tangent to S_2, the magnetic flux through the original surface S is equivalent to the flux through the simpler surface S_1:

$$\psi = \int_S \vec{B} \cdot d\vec{s} \qquad \text{Wb}$$

$$= \int_{S_1} \vec{B} \cdot d\vec{s} + \underbrace{\int_{S_2} \vec{B} \cdot d\vec{s}}_{0} + \underbrace{\int_{S_{\text{ends}}} \vec{B} \cdot d\vec{s}}_{0}$$

$$= \int_{z=0}^{1} \int_{r=R_1}^{R_2} \frac{\mu_0 I}{2\pi r} \underbrace{dr\, dz}_{ds}$$

$$= \frac{\mu_0 I}{2\pi} \ln \frac{R_2}{R_1} \quad \frac{\text{Wb}}{\text{m}} \qquad R_2 > R_1$$

(2.19)

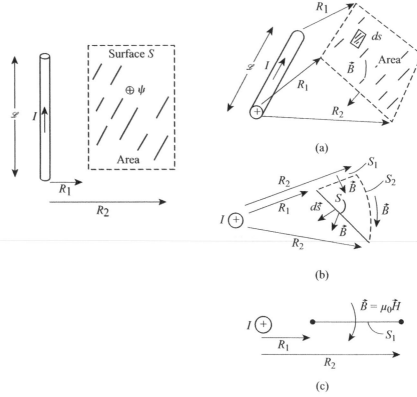

FIGURE 2.7. Determining the dc magnetic flux through a parallel surface of an isolated wire using Gauss's law.

This is a fundamental subproblem for determining the per-unit-length inductance of a parallel-wire line.

Another fundamental subproblem is determining the per-unit-length capacitance of a parallel-wire line. First is determining the dc electric field of an infinitely long isolated wire that carries a dc distribution of charge q C/m that is distributed uniformly along its length and around its periphery, as shown in Fig. 2.8(a).

Constructing a circular contour of radius r outside the wire as shown in Fig. 2.8(b) and assuming that the charge is uniformly distributed along and around the wire and using Gauss's law gives

$$\oint_S \varepsilon \vec{E} \cdot d\vec{s} = Q_{\text{enclosed}} \tag{2.20}$$

where $\varepsilon = \varepsilon_r \varepsilon_0$ and $\varepsilon_0 \equiv (1/36\pi) \times 10^{-9}$ F/m is the permeability of free space and ε_r is the relative (to free space) permittivity of the surrounding

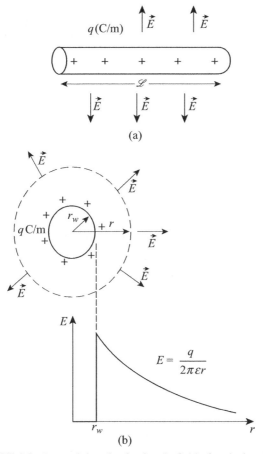

FIGURE 2.8. Determining the dc electric field of an isolated wire.

homogeneous medium. By symmetry we can determine the electric field at a distance r from the wire axis as

$$\oint_s \varepsilon \vec{E} \cdot d\vec{s} = \varepsilon E 2 \pi r \mathcal{L}$$
$$= Q_{\text{enclosed}} \qquad (2.21)$$
$$= q \mathcal{L}$$

thereby giving the electric field directed radially away from the wire as

$$E = \begin{cases} \dfrac{q}{2\pi\varepsilon r} & \dfrac{\text{V}}{\text{m}} & r > r_{\text{w}} \\ 0 & & r < r_{\text{w}} \end{cases} \qquad (2.22)$$

This is plotted in Fig. 2.8(b).

Next we determine the voltage between two points located at radial distances $R_2 > R_1$ from the axis of the wire, as shown in Fig. 2.9. We assume

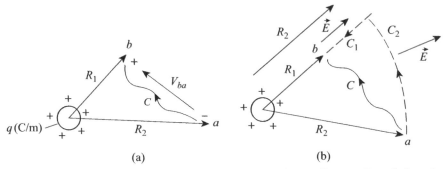

FIGURE 2.9. Determining the voltage between two points that are distances $R_2 > R_1$ from the axis of the wire.

that the point at distance R_1 that is closer to the wire is at the higher potential. The voltage beween the two points is determined as

$$
\begin{aligned}
V_{ba} &= -\int_c \vec{E} \cdot d\vec{l} \\
&= -\int_{c_1} \vec{E} \cdot d\vec{l} - \underbrace{\int_{c_2} \vec{E} \cdot d\vec{l}}_{0} \\
&= -\int_{r=R_2}^{R_1} \frac{q}{2\pi\varepsilon r}\, dr \\
&= \frac{q}{2\pi\varepsilon} \ln \frac{R_2}{R_1} \quad \text{V} \qquad R_2 > R_1
\end{aligned}
\qquad (2.23)
$$

These two fundamental problems allow us to determine the per-unit-length inductance l and capacitance c of a wire-type line.

First we determine the per-unit-length inductance and capacitance of a two-wire line. Consider two wires of radii r_{w1} and r_{w2} separated by a distance s as shown in Fig. 2.10. Using superposition, we obtain the total magnetic flux penetrating the flat surface between the two wires using the subproblem as shown in Fig. 2.10(a) as

$$
\begin{aligned}
\psi &= \frac{\mu_0 I}{2\pi} \ln \frac{s - r_{w2}}{r_{w1}} + \frac{\mu_0 I}{2\pi} \ln \frac{s - r_{w1}}{r_{w2}} \\
&= \frac{\mu_0 I}{2\pi} \ln \frac{(s - r_{w1})(s - r_{w2})}{r_{w1} r_{w2}} \\
&\cong \frac{\mu_0 I}{2\pi} \ln \frac{s^2}{r_{w1} r_{w2}} \qquad s \gg r_{w1}, r_{w2}
\end{aligned}
\qquad (2.24)
$$

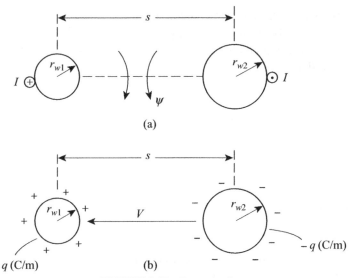

FIGURE 2.10. Two-wire line.

We assume that the wires are of equal radii (as is the practical case) and are "widely separated" sufficiently so that the current and charge are distributed uniformly around the wire peripheries so that the subproblems noted above are satisfied. Since the wires are assumed to be widely separated so that the subproblems are satisfied, this reduces to

$$\psi = \frac{\mu_0 I}{\pi} \ln \frac{s}{r_{\mathrm{w}}} \qquad s \gg r_{\mathrm{w1}} = r_{\mathrm{w2}} \qquad (2.25)$$

Hence the per-unit-length inductance of the line is

$$
\begin{aligned}
l &= \frac{\psi}{I} \\
&= \frac{\mu_0}{\pi} \ln \frac{s}{r_{\mathrm{w}}} \quad \frac{\mathrm{H}}{\mathrm{m}} \qquad s \gg r_{\mathrm{w}}
\end{aligned}
\qquad (2.26)
$$

The per-unit-length capacitance of the line can be obtained from

$$
\begin{aligned}
c &= \frac{q}{2V} \\
&= \frac{1}{v_0^2 l} \\
&= \frac{\mu_0 \varepsilon_0}{l} \\
&= \frac{\pi \varepsilon_0}{\ln(s/r_{\mathrm{w}})} \quad \frac{\mathrm{F}}{\mathrm{m}} \qquad s \gg r_{\mathrm{w}}
\end{aligned}
\qquad (2.27)
$$

If the two wires are "closely spaced," the charges and currents will not be distributed uniformly around the wire peripheries but will concentrate on the facing sides. This is called the *proximity effect*. The exact solution for the per-unit-length inductance is

$$l_{exact} = \frac{\mu_0}{\pi} \ln \left[\frac{s}{2r_w} + \sqrt{\left(\frac{s}{2r_w}\right)^2 - 1} \right] \qquad \frac{H}{m} \qquad (2.28)$$

Figure 2.11 shows the ratio of the wide-separation approximation and the exact value. For a ratio of $s/r_w = 4$, the error is only 5.3%. For this ratio, one wire would just fit between the two, so this is still "pretty close." For the ratio of $s/r_w = 2$ the wires are just touching.

For the case of one wire above an infinite, perfectly conducting ground plane, the "method of images" can be used to replace the ground plane with an equivalent two-wire problem, as shown in Fig. 2.12. All the fields above the ground plane are the same as that with the ground plane removed. The magnetic fields penetrating the loop between the wire and the ground plane are one-half those between the two wires, and therefore

$$l_{\text{ground plane}} = \tfrac{1}{2} l_{\text{two wire}}$$

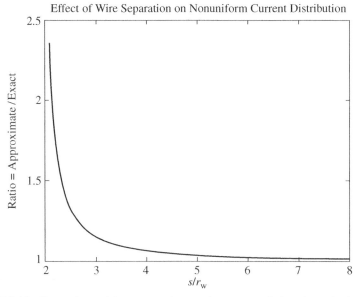

FIGURE 2.11. Comparison of the exact and approximate per-unit-length results for the per-unit-length inductance for a two-wire line.

THE PER-UNIT-LENGTH PARAMETERS   45

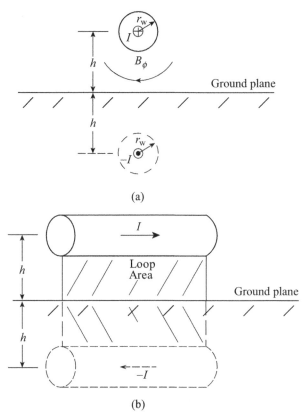

FIGURE 2.12. Transmission line consisting of one wire above an infinite and perfectly conducting ground plane.

Therefore, the per-unit-length inductance for this case is

$$l_{\text{ground plane}} = \frac{\mu_0}{2\pi} \ln\frac{2h}{r_w} \quad \frac{\text{H}}{\text{m}} \qquad h \gg r_w \tag{2.29}$$

This problem can be viewed as being two capacitances in series. Since capacitance in series add like resistors in parallel, we have

$$c_{\text{ground plane}} = 2c_{\text{two wires}}$$
$$= \frac{2\pi\varepsilon_0}{\ln(2h/r_w)} \quad \frac{\text{F}}{\text{m}} \qquad h \gg r_w \tag{2.30}$$

The final type of wire line is the coaxial cable shown in Fig. 2.13. The coaxial cable consists of a wire of radius r_w centered on the axis of an

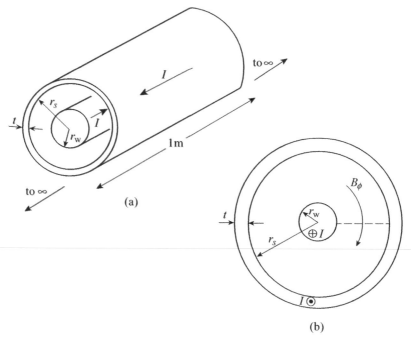

FIGURE 2.13. Coaxial cable.

overall shield of interior radius r_s and thickness t. To determine the per-unit-length inductance l of the cable, we must determine, using the preceding subproblem, the magnetic flux penetrating the flat surface between the surface of the interior wire and the interior surface of the shield as shown in Fig. 2.13(b). By symmetry, the magnetic flux is perpendicular to that surface and we obtain

$$\psi = \int_S \vec{B} \cdot d\vec{s} \qquad \text{Wb}$$

$$= \int_{z=0}^{1} \int_{r=r_w}^{r=r_s} B_\phi \underbrace{dr\,dz}_{ds}$$

$$= \int_{r=r_w}^{r=r_s} \frac{\mu_0 I}{2\pi r} dr$$

$$= \frac{\mu_0 I}{2\pi} \ln \frac{r_s}{r_w} \qquad \frac{\text{Wb}}{\text{m}}$$

(2.31)

Therefore, the per-unit-length inductance is

$$
\begin{aligned}
l &= \frac{\psi}{I} \\
&= \frac{\mu_0}{2\pi} \ln \frac{r_s}{r_w} \qquad \frac{\text{H}}{\text{m}}
\end{aligned}
\tag{2.32}
$$

The per-unit-length capacitance can be found from this as

$$
\begin{aligned}
c &= \frac{\mu_0 \varepsilon_0 \varepsilon_r}{l} \\
&= \frac{2\pi \varepsilon_0 \varepsilon_r}{\ln(r_s/r_w)} \qquad \frac{\text{F}}{\text{m}}
\end{aligned}
\tag{2.33}
$$

where the coaxial cable is filled with a dielectric having a relative permittivity of ε_r. Observe that this result is exact and is not affected by proximity effect.

2.2.2 Lines of Rectangular Cross Section

The per-unit-length parameters of conductors of rectangular cross section (PCB lands) are very difficult to derive. These results are generally obtained by numerical methods and are given in terms of the characteristic impedance of the line and the effective relative permittivity of the line, ε_r'.

The first case is that of a stripline shown in Fig. 2.14. This is typically found on innerplane PCBs, where lands are buried between ground and power planes. The center strip (land) is of width w and is midway between two planes that are separated a distance s. The space between the planes and surrounding

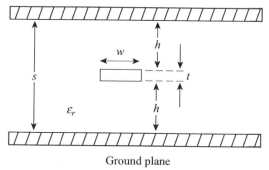

Ground plane

FIGURE 2.14. Stripline.

the strip is filled with a homogeneous dielectric having a relative permittivity ε_r. Assuming a zero thickness strip, $t = 0$, the per-unit-length inductance is

$$l = \frac{30\pi}{v_0} \frac{1}{(w_e/s) + 0.441} \qquad \frac{\text{H}}{\text{m}} \qquad (2.34)$$

and the effective width of the center conductor is

$$\frac{w_e}{s} = \begin{cases} \dfrac{w}{s} & \dfrac{w}{s} \geq 0.35 \\ \dfrac{w}{s} - \left(0.35 - \dfrac{w}{s}\right)^2 & \dfrac{w}{s} \leq 0.35 \end{cases} \qquad (2.35)$$

The per-unit-length capacitance can be found from the inductance since the surrounding medium is homogeneous:

$$c = \frac{\varepsilon_r}{lv_0^2}$$
$$= \frac{\varepsilon_r}{30\pi v_0} \left(\frac{w_e}{s} + 0.441\right) \qquad \frac{\text{F}}{\text{m}} \qquad (2.36)$$

The second configuration is that of a microstrip where a land of width w is situated on top of a substrate that has a relative permittivity of ε_r and a thickness h as shown in Fig. 2.15. This is typical of innerplane PCBs, where lands are present on the top surface of a PCB that has innerplanes buried in it. The substrate is on top of an infinite perfectly conducting ground plane. Assuming a zero-thickness strip, $t = 0$, the per-unit-length inductance is

$$l = \begin{cases} \dfrac{60}{v_0} \ln\left(\dfrac{8h}{w} + \dfrac{w}{4h}\right) & \dfrac{\text{H}}{\text{m}} \quad \dfrac{w}{h} \leq 1 \\ \dfrac{120\pi}{v_0} \left[\dfrac{w}{h} + 1.393 + 0.667\ln\left(\dfrac{w}{h} + 1.444\right)\right]^{-1} & \dfrac{\text{H}}{\text{m}} \quad \dfrac{w}{h} \geq 1 \end{cases} \qquad (2.37)$$

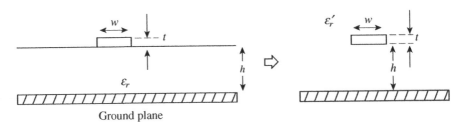

Ground plane

FIGURE 2.15. Microstrip line.

The effective relative permittivity is

$$\varepsilon'_r = \frac{\varepsilon_r + 1}{2} + \frac{\varepsilon_r - 1}{2} \frac{1}{\sqrt{1 + 12(h/w)}} \qquad (2.38)$$

This effective relative permittivity accounts for the fact that the electric field lines are partly in air and partly in the substrate dielectric. If this inhomogeneous medium (air and the dielectric) is replaced by a homogeneous medium having an effective relative permittivity of ε'_r as shown, all properties of the line remain unchanged. But the homogeneous medium problem is much easier to analyze. Hence the per-unit-length capacitance is

$$c = \frac{\varepsilon'_r}{l v_0^2}$$

$$= \begin{cases} \dfrac{\varepsilon'_r}{60 v_0 \ln[(8h/w) + (/4h)]} & \dfrac{\text{F}}{\text{m}} & \dfrac{w}{h} \le 1 \qquad (2.39) \\[3mm] \dfrac{\varepsilon'_r}{120\pi v_0} \left[\dfrac{w}{h} + 1.393 + 0.667\ln\left(\dfrac{w}{h} + 1.444\right)\right] & \dfrac{\text{F}}{\text{m}} & \dfrac{w}{h} \ge 1 \end{cases}$$

The final configuration is that used on low-cost PCBs consisting of two lands on top of a substrate having a relative permittivity of ε_r. The lands are of width w and assumed thickness $t = 0$ and are separated by an edge-to-edge separation s as shown in Fig. 2.16. The per-unit-length inductance is

$$l = \begin{cases} \dfrac{120}{v_0} \ln\left(2\dfrac{1 + \sqrt{k}}{1 - \sqrt{k}}\right) & \dfrac{\text{H}}{\text{m}} & \dfrac{1}{\sqrt{2}} \le k \le 1 \\[3mm] \dfrac{377\pi}{v_0 \ln\left(2\dfrac{1 + \sqrt{k'}}{1 - \sqrt{k'}}\right)} & \dfrac{\text{H}}{\text{m}} & 0 \le k \le \dfrac{1}{\sqrt{2}} \end{cases} \qquad (2.40a)$$

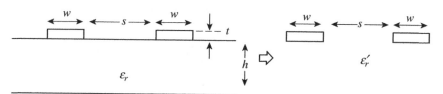

FIGURE 2.16. Printed circuit board.

where k is

$$k = \frac{s}{s + 2w} \qquad (2.40b)$$

and $k' = \sqrt{1 - k^2}$. The effective relative permittivity is

$$\varepsilon_r' = \frac{\varepsilon_r + 1}{2} \left\{ \tanh\left(0.775 \ln\frac{h}{w} + 1.75 \right) \right.$$

$$\left. + \frac{kw}{h}[0.04 - 0.7k + 0.01(1 - 0.1\varepsilon_r)(0.25 + k)] \right\} \qquad (2.41)$$

which again accounts for the fact that the electric field lines are partly in air and partly in the substrate dielectric. If this inhomogeneous medium (air and the dielectric) is replaced by a homogeneous medium having an effective relative permittivity of ε_r', all properties of the line remain unchanged. Hence the per-unit-length capacitance is

$$c = \frac{\varepsilon_r'}{lv_0^2}$$

$$= \begin{cases} \dfrac{\varepsilon_r'}{120 v_0 \ln\left(2\dfrac{1 + \sqrt{k}}{1 - \sqrt{k}} \right)} & \dfrac{F}{m} \quad \dfrac{1}{\sqrt{2}} \le k \le 1 \\[6mm] \dfrac{\varepsilon_r' \ln\left(2\dfrac{1 + \sqrt{k'}}{1 - \sqrt{k'}} \right)}{377\pi v_0} & \dfrac{F}{m} \quad 0 \le k \le \dfrac{1}{\sqrt{2}} \end{cases} \qquad (2.42)$$

2.3 THE GENERAL SOLUTIONS FOR THE LINE VOLTAGE AND CURRENT

The general solutions for the transmission-line voltages and currents are given in terms of traveling waves in (2.11). First look at the load, $z = \mathscr{L}$. The one-way time delay is

$$\boxed{T_D = \frac{\mathscr{L}}{v}} \qquad (2.43)$$

At the load we have $z = \mathscr{L}$:

$$V(\mathscr{L}, t) = V^+(t - T_D) + V^-(t + T_D)$$

$$I(\mathscr{L}, t) = \frac{1}{Z_C} V^+(t - T_D) - \frac{1}{Z_C} V^-(t + T_D) \tag{2.44}$$

For a resistive load of R_L, Ohm's law relates the *total* load voltage and current as

$$\frac{V(\mathscr{L}, t)}{I(\mathscr{L}, t)} = R_L \tag{2.45}$$

Suppose that we have *only* a forward-traveling (incoming) wave at the load:

$$\frac{V(\mathscr{L}, t)}{I(\mathscr{L}, t)} = \frac{V^+(t - T_D)}{(1/Z_C)V^+(t - T_D)} = Z_C \tag{2.46}$$

If the load is *matched* (i.e., $R_L = Z_C$), we will have only a forward-traveling (incoming) wave at the load and there will be no reflected waves at the load if $R_L = Z_C$. But for some general load that is not matched, $R_L \neq Z_C$, we must have an incident (forward-traveling) wave and a reflected (backward-traveling) wave at the load in order to satisfy Ohm's law.

Define the *voltage reflection coefficient at the load* as the ratio of the reflected and incident voltage waves:

$$\boxed{\Gamma_L = \frac{V^-(t + T_D)}{V^+(t - T_D)}} \tag{2.47}$$

If we know the load reflection coefficient, Γ_L, we can determine the reflected voltage wave knowing the incident voltage wave. The *total* voltage and current at the load can then be written in terms of the load reflection coefficient as

$$\boxed{\begin{aligned} V(\mathscr{L}, t) &= V^+(t - T_D)[1 + \Gamma_L] \\ I(\mathscr{L}, t) &= \frac{1}{Z_C} V^+(t - T_D)[1 - \Gamma_L] \end{aligned}} \tag{2.48}$$

Taking the ratio of these two relations gives

$$
\frac{V(\mathscr{L}, t)}{I(\mathscr{L}, t)} = R_L
$$

$$
= Z_C \frac{1 + \Gamma_L}{1 - \Gamma_L} \tag{2.49}
$$

Solving this gives the *voltage reflection coefficient at the load* as

$$
\boxed{\Gamma_L = \frac{R_L - Z_C}{R_L + Z_C}} \tag{2.50}
$$

Observe that since there is a minus sign in the current relation, the current reflection coefficient is the negative of the voltage reflection coefficient:

$$
\boxed{\Gamma_L|_{\text{current}} = -\Gamma_L|_{\text{voltage}}} \tag{2.51}
$$

The process of reflection at the load is like a mirror: the reflected wave is coming out of the mirror and the incident wave is going in as illustrated in Fig. 2.17. *The total voltage is the sum of the incident and reflected waves.*

Now we investigate what happens at the source. The voltage or current wave that was reflected at the load travels back to the source in another time interval of T_D, where it is reflected with a voltage reflection coefficient of

$$
\boxed{\Gamma_S = \frac{R_S - Z_C}{R_S + Z_C}} \tag{2.52}
$$

and sent back to the load. The current reflection coefficient at the source is, again, the negative of the voltage reflection coefficient at the source:

$$
\boxed{\Gamma_S|_{\text{current}} = -\Gamma_S|_{\text{voltage}}} \tag{2.53}
$$

Finally, we obtain the wave sent out initially. We reason that when the source voltage is turned on initially, an initial forward-traveling wave is sent

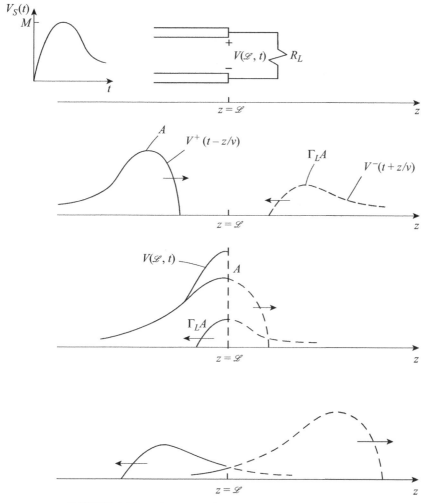

FIGURE 2.17. Reflection at the end of a transmission line.

out toward the load. This initial wave will take a time delay of T_D to get to the load. Any reflections of this initial wave at a mismatched load will require another one-way time delay of T_D to get back to the source. Hence no reflected wave will have arrived at the source over the time interval $0 < t < 2T_D$. So the total voltage at the source is just the forward-traveling wave sent out initially, and hence the ratio of the total voltage to the total current at the source end of the line, $z = 0$, will simply be

$$\frac{V(0,t)}{I(0,t)} = \frac{V^+(t-0)}{(1/Z_C)V^+(t-0)} = Z_C \qquad 0 < t < 2T_D \qquad (2.54)$$

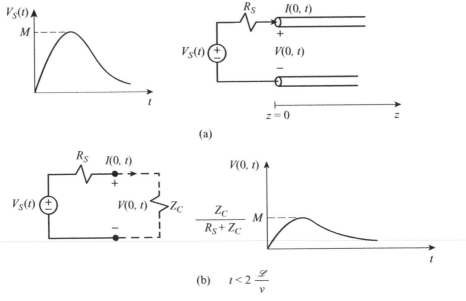

(a)

(b) $t < 2 \dfrac{\mathscr{L}}{v}$

FIGURE 2.18. Input impedance to the line for $0 \le t < 2T_D$.

So the input impedance to the line appears to be Z_C *but only for* $t \le 2T_D$. Hence from Fig. 2.18(b) we can calculate the voltage and current waves sent out initially

$$\boxed{V_{\text{init}} = \frac{Z_C}{R_S + Z_C} V_S(t)}$$ (2.55)

$$\boxed{I_{\text{init}} = \frac{V_S(t)}{R_S + Z_C}}$$ (2.56)

as illustrated in Fig. 2.18.

2.4   WAVE TRACING AND REFLECTION COEFFICIENTS

We trace the waves noting that the total voltage or current at any point on the line and at any time is the sum of the forward- and backward-traveling waves at that point and at that time on the line.

EXAMPLE

Sketch the load voltage, $V(\mathscr{L}, t)$, and the input current to the line, $I(0, t)$, for the problem of Fig. 2.19 at various times that are multiples of the one-way time delay, T_D. First we perform initial computations for the voltage where $V_S(t) = 5u(t)$:

$$T_D = \frac{\mathscr{L}}{v}$$

$$V_{\text{init}} = \frac{Z_C}{R_S + Z_C} V_S(t)$$
$$= \frac{150}{10 + 150} 5$$
$$= 4.688 \text{ V}$$

$$\Gamma_S = \frac{R_S - Z_C}{R_S + Z_C} \qquad \Gamma_L = \frac{R_L - Z_C}{R_L + Z_C}$$
$$= \frac{10 - 150}{10 + 150} \qquad = \frac{\infty - 150}{\infty + 150}$$
$$= -\frac{7}{8} = -0.875 \qquad = +1$$

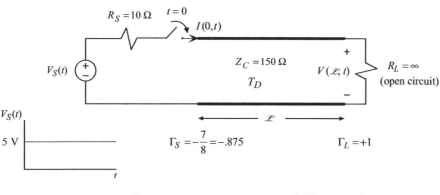

$$V_{\text{init}} = \frac{Z_C}{R_S + Z_C} V_S(t) = 4.688 \text{ V} \qquad I_{\text{init}} = \frac{V_S(t)}{R_S + Z_C} = 31.25 \text{ mA}$$

FIGURE 2.19. An example.

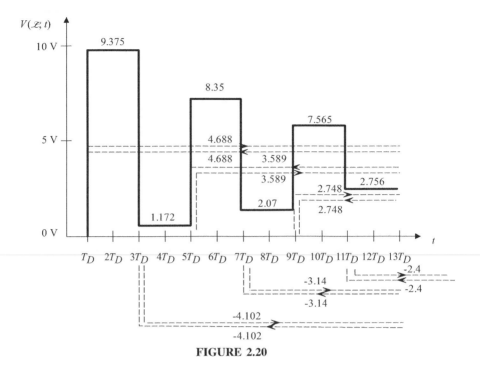

FIGURE 2.20

Next we sketch the load voltage, $V(\mathscr{L}, t)$, versus time, as shown in Fig. 2.20. Next we sketch the input current to the line versus time, $I(0, t)$. First perform the initial computations for the current:

$$I_{\text{init}} = \frac{V_S}{R_S + Z_C}$$

$$= \frac{5}{10 + 150}$$

$$= 31.25 \text{ mA}$$

$$\Gamma_S|_{\text{current}} = -\Gamma_S|_{\text{voltage}}$$

$$= \frac{7}{8}$$

$$\Gamma_L|_{\text{current}} = -\Gamma_L|_{\text{voltage}}$$

$$= -1$$

Next sketch the line input current, $I(0, t)$, versus time as shown in Fig. 2.21. Always verify that your solution is approaching the steady-state values, which for this example are $V(\mathscr{L}, t)_{\text{steady state}} = 5 \text{ V}$ and $I(0, t)_{\text{steady state}} = 0 \text{ A}$.

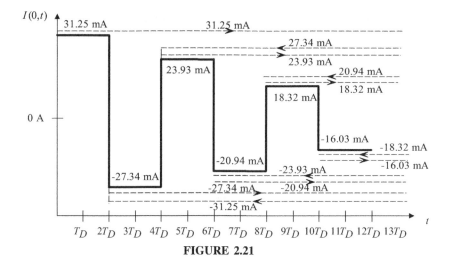

FIGURE 2.21

EXAMPLE

Sketch the voltage at the input, $V(0, t)$, and the current at the output, $I(\mathscr{L}, t)$, of the line versus time for the problem of Fig. 2.22. This problem illustrates the case where the source voltage waveform, $V_S(t)$, is a pulse of finite duration that is several one-way time delays of the line, $T_D = \mathscr{L}/v$, in duration. Hence the

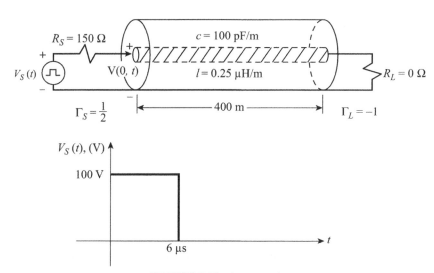

FIGURE 2.22. An example.

incident and reflected pulses from opposite terminations overlap in time and combine to give very complicated total waveshapes at those terminations.

First perform the initial computations:

$$Z_C = \sqrt{\frac{l}{c}}$$

$$= \sqrt{\frac{0.25 \times 10^{-6}}{100 \times 10^{-12}}}$$

$$= 50\,\Omega$$

$$v = \frac{1}{\sqrt{lc}}$$

$$= \frac{1}{\sqrt{(0.25 \times 10^{-6})(100 \times 10^{-12})}}$$

$$= 200\,\text{m}/\mu\text{s}$$

$$T_D = \frac{\mathscr{L}}{v}$$
$$= 2\,\mu\text{s}$$

Perform the initial computations for the voltage:

$$V_{init} = \frac{Z_C}{R_S + Z_C} V_S(t)$$

$$= \frac{50}{150 + 50} 100$$

$$= 25\,\text{V}$$

$$\Gamma_S = \frac{R_S - Z_C}{R_S + Z_C} \qquad \Gamma_L = \frac{R_L - Z_C}{R_L + Z_C}$$

$$= \frac{150 - 50}{150 + 50} \qquad = \frac{0 - 50}{0 + 50}$$

$$= \frac{1}{2} \qquad = -1$$

Sketch the input voltage to the line, $V(0, t)$, as shown in Fig. 2.23.

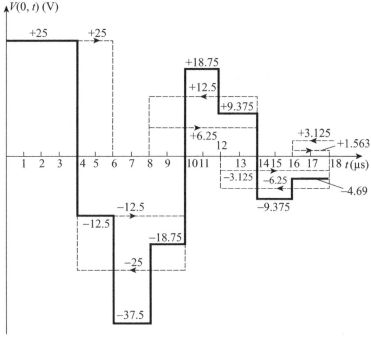

FIGURE 2.23

Verify that the solution is approaching the steady-state value, which for this case is $V(0,t)_{\text{steady state}} = 0$.

Next perform the initial computations for the current:

$$I_{\text{init}} = \frac{1}{R_S + Z_C} V_S(t)$$

$$= \frac{1}{150 + 50} 100$$

$$= 0.5 \text{ A}$$

$$\Gamma_S|_{\text{current}} = -\Gamma_S|_{\text{voltage}} \qquad \Gamma_L|_{\text{current}} = -\Gamma|_{\text{voltage}}$$

$$= -\frac{1}{2} \qquad\qquad\qquad = +1$$

Sketch the load current for the line, $I(\mathscr{L},t)$, as shown in Fig. 2.24. Verify that the solution is approaching the steady-state value, which for this case is $I(\mathscr{L},t)_{\text{steady state}} = 0$.

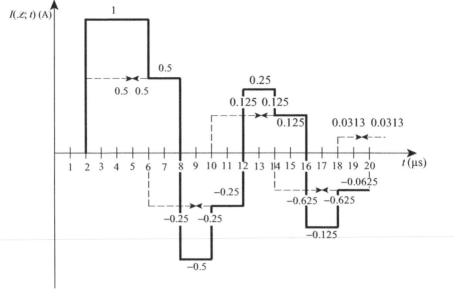

FIGURE 2.24

2.5   A SIMPLE ALTERNATIVE TO WAVE TRACING IN THE SOLUTION OF TRANSMISSION LINES

We can trace the incident and reflected waves giving closed-form solutions for the terminal waveforms in terms of symbols which do NOT have to be repeated for every problem. To obtain these solutions, we use the form of "lattice diagram" shown in Fig. 2.25, which is *normalized* for a unity wave that is launched initially. Time is recorded on the vertical axis in increments of the one-way time delay T_D, and positions along the line are recorded on the horizontal axis. At a time point on the vertical axis where an incident and a reflected wave are present, the incident wave is multiplied by the reflection coefficient and the two waves are added.

The following TOTAL solutions have identical FORMS for ALL problems. For the terminal voltages these are

$$V(0,t) = \frac{Z_C}{R_S + Z_C} V_S(t) + \frac{Z_C}{R_S + Z_C} (1 + \Gamma_S)\Gamma_L \big[V_S(t - 2T_D) \\ + (\Gamma_S \Gamma_L) V_S(t - 4T_D) + (\Gamma_S \Gamma_L)^2 V_S(t - 6T_D) + \cdots \big]$$

(2.57a)

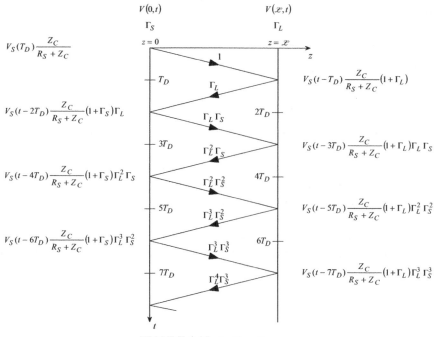

FIGURE 2.25. Lattice diagram.

and

$$V(\mathscr{L},t) = \frac{Z_C}{R_S + Z_C}(1 + \Gamma_L)\left[V_S(t - T_D) + (\Gamma_S\Gamma_L)V_S(t - 3T_D)\right.$$
$$\left. + (\Gamma_S\Gamma_L)^2 V_S(t - 5T_D) + (\Gamma_S\Gamma_L)^3 V_S(t - 7T_D) + \cdots\right]$$

(2.57b)

The terminal current solutions are obtained similarly from the voltage solutions but with the reflection coefficients for currents being the negative of the reflection coefficients for the voltages, as shown in (2.51) and (2.53), and the current wave sent out initially is given by (2.56). Hence the symbolic solutions for the total terminal currents are

$$I(0,t) = \frac{1}{R_S + Z_C}V_S(t) + \frac{1}{R_S + Z_C}(1 - \Gamma_S)(-\Gamma_L)\left[V_S(t - 2T_D)\right.$$
$$\left. + (\Gamma_S\Gamma_L)V_S(t - 4T_D) + (\Gamma_S\Gamma_L)^2 V_S(t - 6T_D) + \cdots\right]$$

(2.58a)

and

$$I(\mathscr{L},t) = \frac{1}{R_S + Z_C}(1 - \Gamma_L)\Big[V_S(t - T_D) + (\Gamma_S\Gamma_L)V_S(t - 3T_D) \\ + (\Gamma_S\Gamma_L)^2 V_S(t - 5T_D) + (\Gamma_S\Gamma_L)^3 V_S(t - 7T_D) + \cdots\Big]$$ (2.58b)

where Γ_S and Γ_L in the current expressions are the voltage reflection coefficients but with their signs in the equations reversed.

Observe in these expressions that the total voltages and currents at the input and the output to the transmission line are combinations of the source waveform, $V_S(t)$, that are delayed by two time delays. Also note that the magnitudes of the source and load reflection coefficients are less than or equal to unity:

$$|\Gamma_S| \leq 1$$
$$|\Gamma_L| \leq 1$$

Observe that the total terminal voltage waveforms, $V(0,t)$ and $V(\mathscr{L},t)$, in (2.57) and total terminal current waveforms, $I(0,t)$ and $I(\mathscr{L},t)$, in (2.58) are sums of delayed replicas of $V_S(t)$ multiplied by products of the source and load reflection coefficients, $(\Gamma_S\Gamma_L)^n$, which are also progressively less than unity. Hence if the source resistor is less than the characteristic impedance, $R_S < Z_C$, and the load resistor is greater than the load resistor, $R_L > Z_C$, or vice versa, the source and load reflection coefficients are of *opposite sign*. Hence a resulting terminal voltage will have a portion added to it and subtracted from it, resulting in *oscillations*. On the other hand, if the source resistor and the load resistor are both less than the characteristic impedance, $R_S < Z_C$, $R_L < Z_C$, or are both greater than the characteristic impedance, $R_S > Z_C$, $R_L > Z_C$, the source and load reflection coefficients are of the *same sign* and the terminal voltage will steadily build up to its steady-state value. These observations also apply to the terminal currents.

The voltage results in (2.57) are multiplied by a factor representing the voltage division at the source that was used to determine the voltage sent out initially: $Z_C/(R_S + Z_C)$. Similarly, the current results in (2.58) are multiplied by $1/R_S + Z_C$. Finally, each result is multiplied by a constant: $(1 + \Gamma_S)\Gamma_L$ for $V(0,t)$ and $(1 + \Gamma_L)$ for $V(\mathscr{L},t)$ and $(1 - \Gamma_S)(-\Gamma_L)$ for $I(0,t)$ and $(1 - \Gamma_L)$ for $I(\mathscr{L},t)$, where these current coefficients result from negating the voltage reflection coefficients to give the corresponding current reflection coefficients. Once these coefficients are determined, the component waveforms are plotted in terms of the scaled and delayed source waveform $V_S(t)$. Once this is completed, these waveforms are summed to give the total waveforms $V(0, t)$, $V(\mathscr{L}, t)$, $I(0, t)$, and $I(\mathscr{L}, t)$, and the analysis is complete.

EXAMPLE

Determine the load voltage, $V(\mathscr{L}, t)$, and the input current, $I(0, t)$, for the circuit determined by wave tracing for the circuit in Fig. 2.19, where $V_S(t) = 5u(t)$, $R_S = 10$, $R_L = \infty$, and $Z_C = 150$.
  From (2.57b) and (2.58a) we obtain

$$V(\mathscr{L}, t) = \frac{Z_C}{R_S + Z_C}(1 + \Gamma_L)\big[V_S(t - T_D) + (\Gamma_S\Gamma_L)V_S(t - 3T_D)$$
$$+(\Gamma_S\Gamma_L)^2 V_S(t - 5T_D) + (\Gamma_S\Gamma_L)^3 V_S(t - 7T_D) + \cdots\big]$$
(2.57b)

and

$$I(0, t) = \frac{1}{R_S + Z_C}V_S(t) + \frac{1}{R_S + Z_C}(1 - \Gamma_S)(-\Gamma_L)\big[V_S(t - 2T_D)$$
$$+(\Gamma_S\Gamma_L)V_S(t - 4T_D) + (\Gamma_S\Gamma_L)^2 V_S(t - 6T_D) + \cdots\big]$$
(2.58a)

For this problem we calculate for the voltage $Z_C/(R_S + Z_C) = 15/16 = 0.9375$, $\Gamma_S = -0.875$, $\Gamma_L = 1$, $1 + \Gamma_L = 1 + 1 = 2$, and $\Gamma_S\Gamma_L = -0.875$. We obtain

$$V(\mathscr{L}, t) = 1.875\big[V_S(t - T_D) - 0.875V_S(t - 3T_D)$$
$$+0.7656V_S(t - 5T_D) - 0.6699V_S(t - 7T_D) + \cdots\big]$$
$$= 1.875\big[V_S(t - T_D) - 1.641V_S(t - 3T_D)$$
$$+1.436V_S(t - 5T_D) - 1.256V_S(t - 7T_D) + \cdots\big]$$
(2.57b)

Plotting these terms in Fig. 2.26 and summing at each point in time gives the result obtained by wave tracing in Fig. 2.20.
  For the current we obtain

$$\frac{1}{R_S + Z_C} = \frac{1}{160} = 6.25 \times 10^{-3}$$

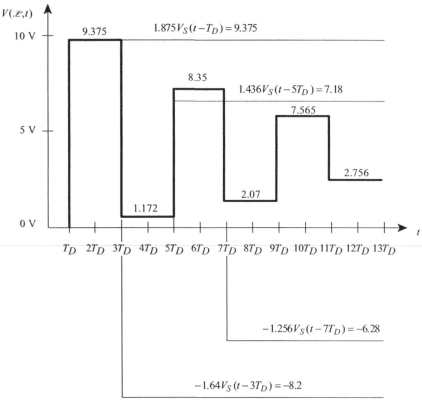

FIGURE 2.26. Example in Fig. 2.20.

and

$$(1 - \Gamma_S)(-\Gamma_L) = \left(1 - \frac{10 - 150}{10 + 150}\right)(-1) = -\frac{15}{8} = -1.875$$

and $(\Gamma_S\Gamma_L) = -0.875$. We obtain

$$
\begin{aligned}
I(0, t) &= 6.25 \times 10^{-3} V_S(t) - 11.72 \times 10^{-3} \left[V_S(t - 2T_D)\right. \\
&\quad \left. -0.875 V_S(t - 4T_D) + 0.7656 V_S(t - 6T_D) + \cdots\right] \\
&= 6.25 \times 10^{-3} V_S(t) - 11.72 \times 10^{-3} [V_S(t - 2T_D) \\
&\quad + 10.26 \times 10^{-3} V_S(t - 4T_D) - 8.972 \times 10^{-3} V_S(t - 6T_D) + \cdots]
\end{aligned}
$$

$$(2.58a)$$

Plotting these terms in Fig. 2.27 and summing at each point in time gives the result obtained by wave tracing in Fig. 2.21.

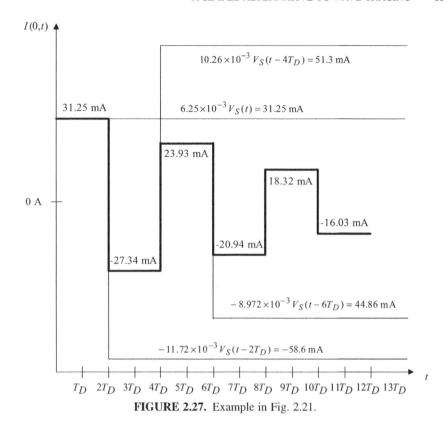

FIGURE 2.27. Example in Fig. 2.21.

EXAMPLE

Sketch the input voltage, $V(0, t)$, and load current, $I(\mathscr{L}, t)$, in Fig. 2.22 using this method.

For this example we compute the factors

$$\frac{Z_C}{R_S + Z_C} = \frac{1}{4} \quad \text{and} \quad (1 + \Gamma_S)\Gamma_L = -\frac{3}{2}$$

for $V(0, t)$ and $(1 + \Gamma_L) = 0$ for $V(\mathscr{L}, t)$. For the plot of $V(0, t))$ the series expression in (2.57a) becomes

$$V(0, t) = \frac{1}{4}V_S(t) - \frac{3}{8}\left[V_S(t - 2T_D) - \frac{1}{2}V_S(t - 4T_D) + \frac{1}{4}V_S(t - 6T_D)\right.$$
$$\left. - \frac{1}{8}V_S(t - 8T_D) + \cdots\right]$$

$$=\frac{1}{4}V_S(t)-\frac{3}{8}\left[V_S(t-2T_D)+\frac{3}{16}V_S(t-4T_D)-\frac{3}{32}V_S(t-6T_D)\right.$$

$$\left.+\frac{3}{64}V_S(t-8T_D)+\cdots\right]$$

which is shown in Fig. 2.28. Adding the pulses gives the solution for $V(0,t)$ shown in Fig. 2.29.

The series expression for $I(\mathscr{L},t)$ in (2.58b) becomes

$$I(\mathscr{L},t)=\frac{1}{200}2\left[V_S(t-T_D)-\frac{1}{2}V_S(t-3T_D)+\frac{1}{4}V_S(t-5T_D)\right.$$

$$\left.-\frac{1}{8}V_S(t-7T_D)+\cdots\right]$$

$$=\frac{1}{100}\left[V_S(t-T_D)-\frac{1}{200}V_S(t-3T_D)+\frac{1}{400}V_S(t-5T_D)\right.$$

$$\left.-\frac{1}{800}V_S(t-7T_D)+\cdots\right]$$

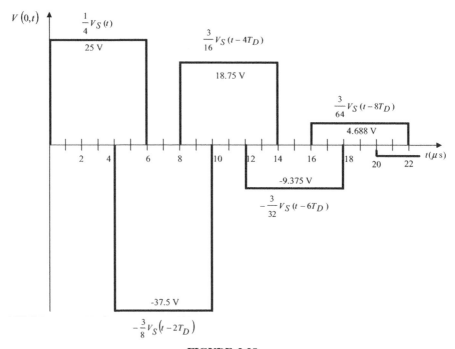

FIGURE 2.28

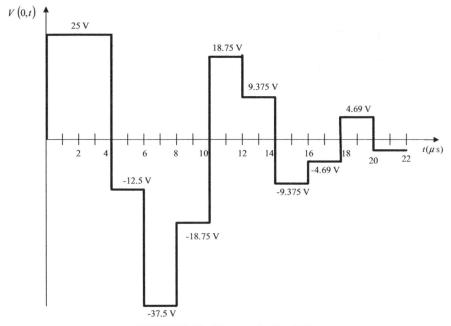

FIGURE 2.29. Example in Fig. 2.23.

The coefficients are $1/(R_S + Z_C) = \frac{1}{200}$, $(1 + -\Gamma_S)(-\Gamma_L) = \frac{1}{2}$ for $I(0,t)$, and $(1 + -\Gamma_L) = 2$ for $I(\mathscr{L},t)$. Figure 2.30 shows this summation in terms of the source pulse, $V_S(t)$. Adding the pulses gives the solution for $I(\mathscr{L},t)$), shown in Fig. 2.31.

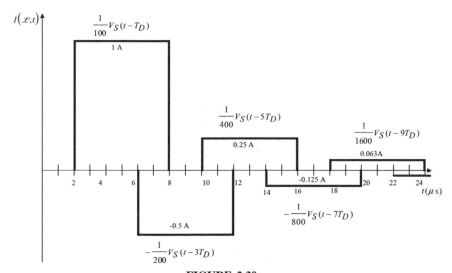

FIGURE 2.30

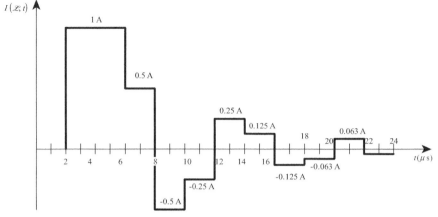

FIGURE 2.31. Example in Fig. 2.24.

EXAMPLE

As another example and one in which $V_S(t)$ is complicated, suppose that $V_S(t)$ is again a pulse of 100 V and duration of 6 μs but steadily ramps from 0 V at $t = 0$ s to 100 V at $t = 6$ μs, at which time it goes to zero, as shown in Fig. 2.32.

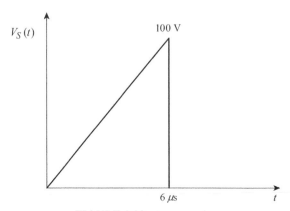

FIGURE 2.32. An example.

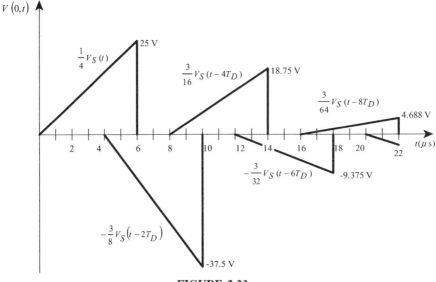

FIGURE 2.33

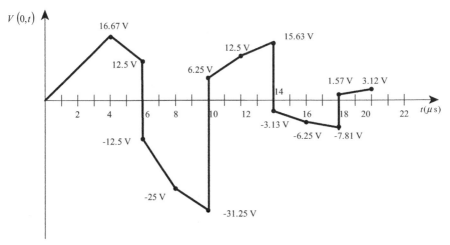

FIGURE 2.34

Since none of the parameters except the waveform for $V_S(t)$ has been changed, the equation for $V(0, t)$ is *unchanged*:

$$V(0,t)=\frac{1}{4}V_S(t) -\frac{3}{8}\left[V_S(t-2T_D) -\frac{1}{2}V_S(t-4T_D) +\frac{1}{4}V_S(t-6T_D)\right.$$

$$\left.-\frac{1}{8}V_S(t-8T_D)+\cdots\right]$$

$$= \frac{1}{4} V_S(t) - \frac{3}{8} V_S(t - 2T_D) + \frac{3}{16} V_S(t - 4T_D) - \frac{3}{32} V_S(t - 6T_D)$$

$$+ \frac{3}{64} V_S(t - 8T_D) + \cdots$$

Plotting the individual components is shown in Fig. 2.33. Adding the pulses gives the solution for $V(0, t)$ shown in Fig. 2.34.

2.6   THE SPICE (PSPICE) EXACT TRANSMISSION-LINE MODEL

SPICE (PSPICE) contains an *exact* model for a *lossless two-conductor transmission line*. The advantage of this model is that (1) it is simple and fast to use, and (2) you can use *any terminations*, inductors, capacitors, or nonlinear resistors such as diodes and transistors and see the exact solution for the voltages and currents versus time at the ends of the line!

SPICE (PSPICE) implements the exact transmission-line solution in a discrete "bootstrapping" manner. Consider the exact solution of the transmission-line equations:

$$
\begin{aligned}
V(z, t) &= V^+\left(t - \frac{z}{v}\right) + V^-\left(t + \frac{z}{v}\right) \\
Z_C I(z, t) &= V^+\left(t - \frac{z}{v}\right) - V^-\left(t + \frac{z}{v}\right)
\end{aligned}
\tag{2.59}
$$

Evaluate them at the source end, $z = 0$:

$$
\begin{aligned}
V(0, t) &= V^+(t) + V^-(t) \\
Z_C I(0, t) &= V^+(t) - V^-(t)
\end{aligned}
\tag{2.60}
$$

Evaluate them at the load end, $z = \mathscr{L}$:

$$
\begin{aligned}
V(\mathscr{L}, t) &= V^+(t - T_D) + V^-(t + T_D) \\
Z_C I(\mathscr{L}, t) &= V^+(t - T_D) - V^-(t + T_D)
\end{aligned}
\tag{2.61}
$$

Add and subtract them:

$$
\begin{aligned}
V(0, t) + Z_C I(0, t) &= 2V^+(t) \\
V(0, t) - Z_C I(0, t) &= 2V^-(t)
\end{aligned}
\tag{2.62a}
$$

$$V(\mathscr{L},t) + Z_C I(\mathscr{L},t) = 2V^+(t - T_D)$$

$$V(\mathscr{L},t) - Z_C I(\mathscr{L},t) = 2V^-(t + T_D)$$

(2.62b)

Time-shift and rearrange:

$$V(0,t) = Z_C I(0,t) + 2V^-(t)$$

$$V(\mathscr{L},t) = -Z_C I(\mathscr{L},t) + 2V^+(t - T_D)$$

(2.63a)

$$V(0,t - T_D) + Z_C I(0,t - T_D) = 2V^+(t - T_D)$$

$$V(\mathscr{L},t - T_D) - Z_C I(\mathscr{L},t - T_D) = 2V^-(t)$$

(2.63b)

This gives

$$V(0,t) = Z_C I(0,t) + \underbrace{[V(\mathscr{L},t - T_D) - Z_C I(\mathscr{L},t - T_D)]}_{E_0(\mathscr{L},t-T_D)}$$

$$V(\mathscr{L},t) = -Z_C I(\mathscr{L},t) + \underbrace{[V(0,t - T_D) + Z_C I(0,t - T_D)]}_{E_\mathscr{L}(0,t-T_D)}$$

(2.64)

This gives the equivalent circuit shown in Fig. 2.35.

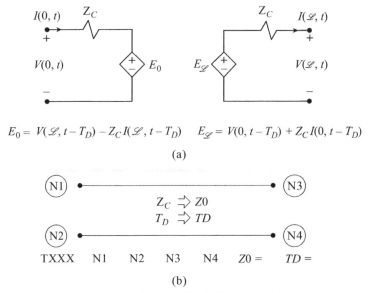

$$E_0 = V(\mathscr{L}, t - T_D) - Z_C I(\mathscr{L}, t - T_D) \qquad E_\mathscr{L} = V(0, t - T_D) + Z_C I(0, t - T_D)$$

(a)

$$Z_C \Rightarrow Z0$$
$$T_D \Rightarrow TD$$

TXXX   N1   N2   N3   N4   Z0 =   TD =

(b)

FIGURE 2.35. SPICE (PSPICE) model.

SPICE (PSPICE) uses a "bootstrapping" or iterative method of solution. On the TRAN line you can set a *minimum step size*, Δt. Then SPICE solves at $t = 0$, uses that solution to determine the solution at $t = \Delta t$, uses that solution to determine the solution at $t = 2\Delta t$, uses that solution to determine the solution at $t = 3\Delta t$, and so on, until it gets to the *final solution time* you specified on the. TRAN line.

We will solve the earlier examples using SPICE (PSPICE).

EXAMPLE

Solve the problem of Fig. 2.19 using PSPICE.

Labeling the nodes of the circuit as shown in Fig. 2.36 gives the PSPICE program:

```
EXAMPLE
VS 1 0 PWL(0 0 0.01N 5)
RS 1 2 10
T 2 0 3 0 Z0=150 TD=1N
RL 3 0 1E8
.TRAN 0.01N 20N 0 0.01N
.PRINT TRAN V(3) I(RS)
.PROBE
.END
```

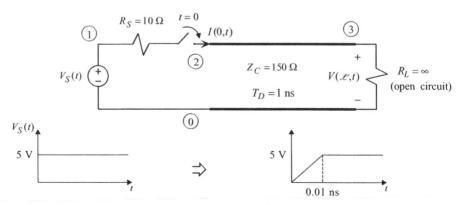

FIGURE 2.36. SPICE model of Fig. 2.19.

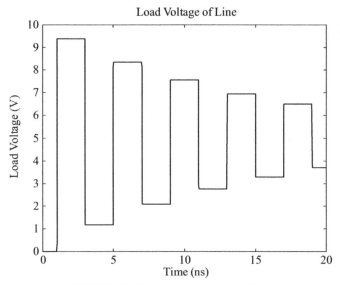

FIGURE 2.37. Load voltage of the line.

Figure 2.37 shows the terminal voltage, $V(\mathscr{L}, t)$, and Figure 2.38 shows the current at the input to the line, $I(0, t)$.

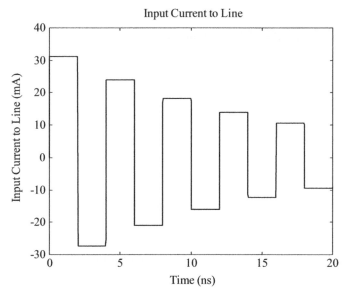

FIGURE 2.38. Input current to the line.

EXAMPLE

Solve the example of Fig. 2.22 using PSPICE.

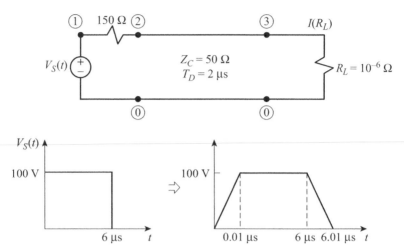

FIGURE 2.39. SPICE model of Fig. 2.22.

See Fig. 2.39. The PSPICE program is

```
EXAMPLE
VS 1 0 PWL(0 0.01U 100 6U 100 6.01U 0)
RS 1 2 150
T 2 0 3 0 Z0=50 TD=2U
RL 3 0 1E-6
.TRAN.01U 20U 0.01U
.PRINT TRAN V(2) I(RL)
.PROBE
.END
```

The results are shown in Fig. 2.40. Check with those obtained by hand.

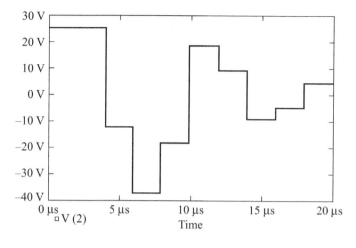

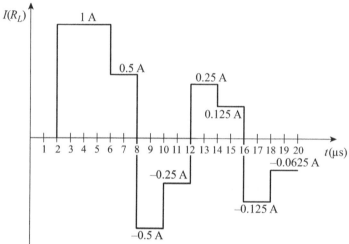

FIGURE 2.40

2.7   LUMPED-CIRCUIT APPROXIMATE MODELS OF THE LINE

A lumped-equivalent circuit model of the transmission line, such as a lumped-pi circuit shown in Fig. 2.41, can provide adequate approximate solutions of the transmission-line equations as long as the line is electrically short at the "significant frequencies" of the source waveform $V_S(t)$ within its bandwidth:

$$\text{BW} = f_{\max} \cong \frac{1}{\tau_r} \qquad \tau_r = \tau_f \qquad (2.65)$$

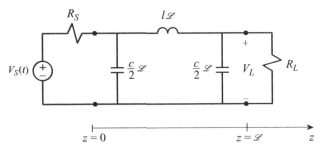

FIGURE 2.41. Lumped-pi equivalent circuit.

In this circuit, the total line inductance, $L = l\mathscr{L}$, is placed in the middle, and the total line capacitance, $C = c\mathscr{L}$, is split and placed on both sides so as to make the model symmetrical, as is the actual line. The criterion for its adequacy is that the total line length must be "electrically short" at this highest significant frequency:

$$\mathscr{L} < \frac{1}{10}\lambda$$
$$= \frac{1}{10}\frac{v}{f_{\max}} \tag{2.66}$$

Substituting gives

$$\boxed{\tau_r > 10T_D} \tag{2.67}$$

But use of this equivalent circuit is unnecessary since an *exact* model of a *lossless* transmission line exists in the PSPICE program.

Since $Z_C = vl$ and $Z_C = 1/vc$ the total inductance and capacitance of the line can be computed in terms of the line characteristic impedance Z_C and the line one-way delay T_D as

$$\boxed{\begin{aligned} L &= l\mathscr{L} \\ &= \frac{Z_C}{v}\mathscr{L} \\ &= Z_C T_D \quad \text{H} \end{aligned}} \tag{2.68a}$$

and

$$\boxed{\begin{aligned} C &= c\mathscr{L} \\ &= \frac{\mathscr{L}}{vZ_C} \\ &= \frac{T_D}{Z_C} \quad \text{F} \end{aligned}} \tag{2.68b}$$

EXAMPLE

As an example we compute the response of a stripline whose cross section is shown in Fig. 2.42(a) and whose terminations are shown in Fig. 2.42(b). It is driven by the 5-V, 50-MHz clock waveform having 500-ps rise and fall times and a 50% duty cycle shown in Fig. 2.42(c).

The load voltage after three cycles of $V_S(t)$ is computed with the PSPICE program below and is shown in Fig. 2.43. The PSPICE program is

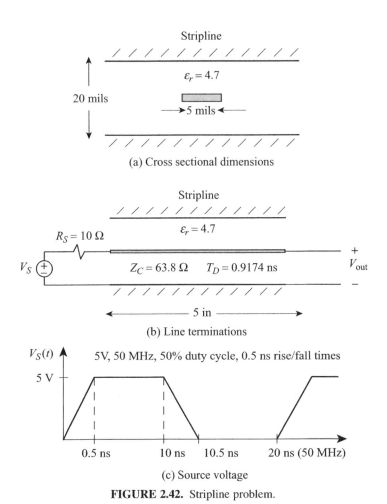

(a) Cross sectional dimensions

(b) Line terminations

(c) Source voltage

FIGURE 2.42. Stripline problem.

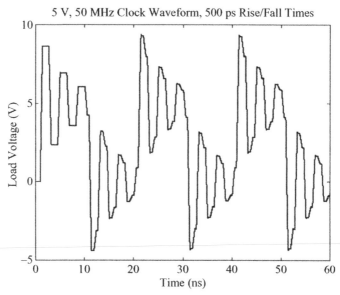

FIGURE 2.43. Load voltage.

```
EXAMPLE
VS 1 0 PULSE(0 5 0 500P 500P 9.5N 20N)
RS 1 2 10
T 2 0 3 0 Z0=63.816 TD=0.9174N
RL 3 0 1E8
.TRAN 10P 60N 0 10P
.PRINT TRAN V(3)
.PROBE
.END
```

Notice that the response to the initial pulse $0 < t < 10.5$ ns has not gotten into steady state when the pulse turns off and remains at zero for 10.5 ns $< t < 20$ ns. But reflections due to that initial half of the cycle continue for 10.5 ns $< t < 20$ ns even though the pulse waveform is turned off. After the first cycle of $V_S(t)$, the load voltage essentially reaches steady state.

EXAMPLE

For the stripline example shown in Fig. 2.42, we can model the line with lumped-pi approximate models as shown in Fig. 2.44 as

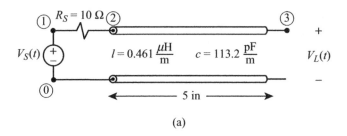

(a)

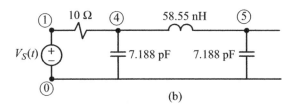

(b)

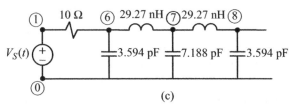

(c)

FIGURE 2.44. Lumped-circuit approximate models of the stripline example in Fig. 2.42.

$l = Z_C/v = 0.461\ \mu\text{H/m}$ and $c = 1/Z_C v = 113.2\ \text{pF/m}$ and totals of $L = Z_C T_D = 58.55\ \text{nH}$ and $C = T_D/Z_C = 14.38\ \text{pF}$. The PSPICE program is

```
EXAMPLE
VS 1 0 PULSE(0 5 0 500P 500P 9.5N 20N)
*VS 1 0 AC 1 0
* TRANSMISSION LINE MODEL
RS1 1 2 10
T 2 0 3 0 Z0=63.816 TD=0.9174N
RL1 3 0 1E8
* ONE-PI SECTION
RS2 1 4 10
C11 4 0 7.188P
L11 4 5 58.55N
C12 5 0 7.188P
RL2 5 0 1E8
* TWO-PI SECTIONS
RS3 1 6 10
C21 6 0 3.594P
```

```
L21 6 7 29.27N
C22 7 0 7.188P
L22 7 8 29.27N
C23 8 0 3.594P
RL3 8 0 1E8
.TRAN 10P 20N 0 10P
.PRINT TRAN V(3) V(5) V(8)
.PROBE
*.AC DEC 100 1E7 1E9
*.PRINT AC VM(3) VM(5) VM(8)
.END
```

Using PSPICE to solve these gives the time-domain solution shown in Fig. 2.45. The line is one wavelength long at $1/T_D = 1.09$ GHz and is electrically short at 109 MHz. Hence the line is electrically short only at the fundamental frequency of the pulse train of 50 MHz. Hence it is not surprising that these lumped-circuit models do a poor job of representing the line response since the bandwidth of the waveform is BW $= 1/\tau_r = 2$ GHz.

If we slow the rise and fall times to 2 ns, the bandwidth of the waveform is reduced to BW $= 1/\tau_r = 500$ MHz. Figure 2.46 shows the time-domain response comparing the exact transmission-line model, one pi section, and two pi sections. The two-pi section model provides a good correlation, but the one-pi model is considerably in error.

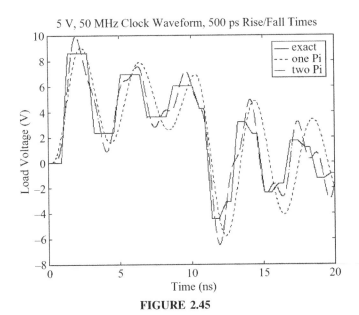

FIGURE 2.45

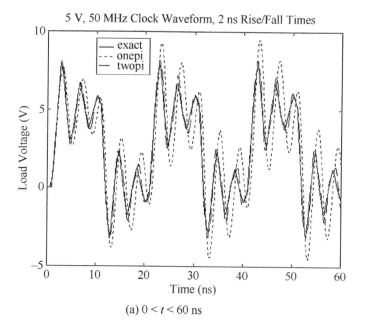

(a) $0 < t < 60$ ns

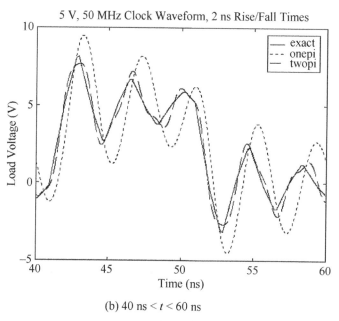

(b) 40 ns $< t < 60$ ns

FIGURE 2.46

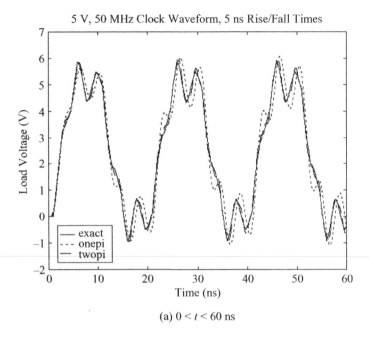

(a) $0 < t < 60$ ns

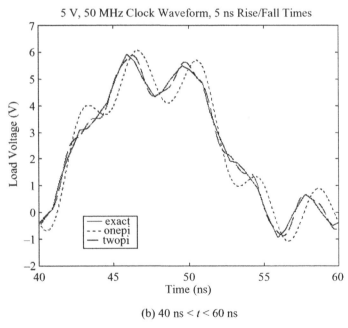

(b) 40 ns $< t < 60$ ns

FIGURE 2.47

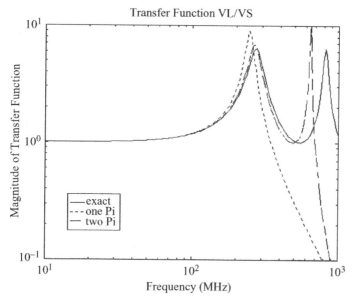

FIGURE 2.48. Frequency domain transfer function.

If we further slow the rise and fall times to 5 ns, the bandwidth of the waveform is reduced to $BW = 1/\tau_r = 200$ MHz. Figure 2.47 shows the time-domain response comparing the exact transmission-line model, one pi section, and two pi sections. The one-pi-section model gives better predictions, and the two-pi-section model provides a closer correlation.

We can easily show the frequency-domain transfer function of the line by changing a few lines of the PSPICE program (those noted with an asterisk as comment lines). The *frequency-domain* response is the response to a single-frequency sinusoidal source, $V_S(t) = V_S \cos(\omega t + \theta_S)$. The load voltage is similar, as $V_L(t) = V_L \cos(\omega t + \theta_L)$. The magnitude of the frequency response is computed as shown in Fig. 2.48. Observe that the one-pi-section model gives adequate predictions compared to the exact transmission-line model up to about 200 MHz, while the two-pi-section model gives adequate predictions up to about 500 MHz. This explains why the response to the 2-ns rise- and fall-time pulse having a bandwidth of 500 MHz is adequately predicted only by the two-pi-section model, but the response to the 5-ns rise- and fall-time pulse having a bandwidth of 200 MHz is reasonably predicted by both the two-pi-section model and the one-pi-section model.

This frequency-response plot shows that using more than one lumped-pi section to try to extend the frequency range doesn't gain very much and results in a larger and larger lumped-circuit model for PSPICE to solve. So if the line is electrically short, a single-section lumped pi model works fine. If the line is NOT electrically short, you have no other choice but to solve the transmission-line equations.

2.8   EFFECTS OF REACTIVE TERMINATIONS ON TERMINAL WAVEFORMS

So far we have sketched the terminal voltages and currents vs. time for resistive terminations. Sketching the corresponding terminal waveforms for reactive and/or nonlinear terminations is very difficult and is unnecessary since we can use the exact model of the transmission line that exists in PSPICE. In this section we briefly investigate the effect of capacitive and inductive terminations.

Hand sketching the terminal voltages and currents for resistive terminations is easy. But for *reactive terminations* such as capacitive and inductive loads, hand sketching the terminal voltages and currents is more difficult. The Laplace transform of the transmission-line equations makes this easier. Transforming the transmission-line equations *with respect to time t* gives

$$\frac{dV(z,s)}{dz} = -slI(z,s)$$

$$\frac{dI(z,s)}{dz} = -scV(z,s)$$

$$(2.69)$$

The Laplace transform converts *partial differential equations* into *ordinary differential equations*. The second-order uncoupled equations are

$$\frac{d^2V(z,s)}{dz^2} - s^2lcV(z,s) = 0$$

$$\frac{d^2I(z,s)}{dz^2} - s^2lcI(z,s) = 0$$

$$(2.70)$$

The solutions are

$$V(z,s) = V^+(s)e^{-s(z/v)} + V^-(s)e^{s(z/v)}$$

$$I(z,s) = \frac{1}{Z_C}V^+(s)e^{-s(z/v)} - \frac{1}{Z_C}V^-(s)e^{s(z/v)}$$

$$(2.71a)$$

where $V^+(s)$ and $V^-(s)$ are to be determined by the source waveform and the terminations. Note that the inverse transform is

$$V(z,t) = V^+\left(t - \frac{z}{v}\right) + V^-\left(t + \frac{z}{v}\right)$$
$$I(z,t) = \frac{1}{Z_C}V^+\left(t - \frac{z}{v}\right) - \frac{1}{Z_C}V^-\left(t + \frac{z}{v}\right)$$

(2.71b)

2.8.1 Effect of Capacitive Terminations

Now we investigate the computation of the terminal voltage of the transmission line when the termination is a capacitor, as illustrated in Fig. 2.49. This solution is usually obtained more easily *indirectly*. The Laplace-transformed circuit is shown in Fig. 2.50. We assumed that the source is matched to make the solution easier, so that $\Gamma_S(s) = 0$. The transformed load reflection coefficient is

$$\Gamma_L(s) = \frac{Z_L(s) - Z_C}{Z_L(s) + Z_C}$$
$$= \frac{(1/sC) - Z_C}{(1/sC) + Z_C}$$
$$= \frac{1 - sT_C}{1 + sT_C}$$

(2.72)

where the time constant is

$$T_C = Z_C C$$

(2.73)

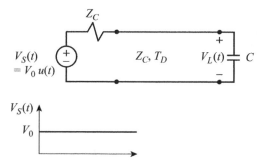

FIGURE 2.49. Termination of a transmission line in a capacitor.

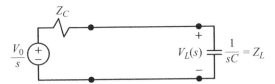

FIGURE 2.50. Laplace-transformed circuit.

Since the source is matched, there will be no reflections there, so we obtain the load voltage as

$$V_L(t) = (1 + \Gamma_L) \frac{Z_C}{(R_S = Z_C) + Z_C} V_0 u(t - T_D) \qquad (2.74)$$

where $u(t)$ is the unit step function. Transforming this yields

$$V_L(s) = (1 + \Gamma_L(s)) \frac{1}{2} \frac{V_0}{s} e^{-sT_D}$$

$$= \frac{1/T_C}{(s + 1/T_C)s} V_0 e^{-sT_D} \qquad (2.75)$$

$$= \left(\frac{1}{s} - \frac{1}{s + 1/T_C} \right) V_0 e^{-sT_D}$$

The inverse transform of this is

$$V_L(t) = V_0 u(t - T_D) - e^{-(t-T_D)/T_C} V_0 u(t - T_D) \qquad (2.76)$$

which is sketched in Fig. 2.51. This makes sense, of course, because when the unit step voltage arrives at the capacitive load, the capacitor initially looks like a short circuit and then begins charging up to V_0. Notice that in addition to the

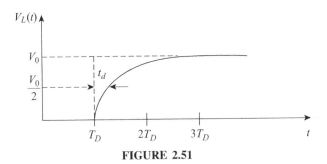

FIGURE 2.51

one-way time delay T_D, there is an additional time delay (to get to half the level) of $V_0 e^{-t/T_C} = \frac{1}{2} V_0$, or

$$t_d = 0.693 T_C \tag{2.77}$$

which is due to the time constant

$$T_C = Z_C C \tag{2.78}$$

Since the line is matched at the source, at the load it looks to the capacitor like a Thévenin equivalent with a source resistance of Z_C.

2.8.2   Effect of Inductive Terminations

Now we investigate in a similar fashion the effect of an inductive load. The Laplace-transformed circuit is shown in Fig. 2.52. Again, since the source is assumed matched,

$$\Gamma_S(s) = 0 \tag{2.79}$$

The transformed load reflection coefficient is

$$\begin{aligned}
\Gamma_L(s) &= \frac{Z_L(s) - Z_C}{Z_L(s) + Z_C} \\
&= \frac{sL - Z_C}{sL + Z_C} \\
&= \frac{sT_L - 1}{sT_L + 1}
\end{aligned} \tag{2.80}$$

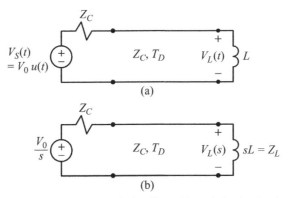

FIGURE 2.52. Transmission line with an inductive load.

where the time constant is

$$T_L = \frac{L}{Z_C} \qquad (2.81)$$

Since the source is matched, there will be no reflections there, so we obtain the load voltage as

$$V_L(t) = (1 + \Gamma_L) \frac{Z_C}{(R_S = Z_C) + Z_C} V_0 u(t - T_D) \qquad (2.82)$$

where $u(t)$ is the unit step function. Transforming this yields

$$V_L(s) = (1 + \Gamma_L(s)) \frac{1}{2} \frac{V_0}{s} e^{-sT_D}$$
$$= \frac{1}{s + 1/T_L} V_0 e^{-sT_D} \qquad (2.83)$$

The inverse transform of this is

$$V_L(t) = e^{-(t-T_D)/T_L} V_0 u(t - T_D) \qquad (2.84)$$

which is sketched in Fig. 2.53. This makes sense, of course, because when the unit step voltage arrives at the inductive load, the inductor initially looks like an open circuit and then transitions to a short circuit in the steady state. Since the line is again matched at the source, at the load it again looks to the inductor like a Thévenin equivalent with a source resistance of Z_C. This method can be similarly used to solve mismatched source and load but the details become more complex.

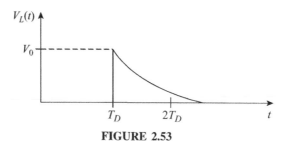

FIGURE 2.53

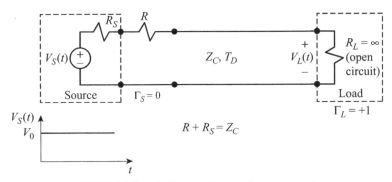

FIGURE 2.54. Series matching at the source end.

2.9   MATCHING SCHEMES FOR SIGNAL INTEGRITY

We have seen how mismatches at terminations on the transmission line can cause ringing and other undesirable waveforms that can cause logic errors in digital systems that are significant problems in high-speed digital systems. The question is: How do we remedy these signal integrity problems?

Series Matching at the Source End (Fig. 2.54)   The first simple scheme is *series matching,* where we place an additional resistor in series with the input to the line to bring the total source resistance to Z_C and therefore match the line at its source as illustrated in Fig. 2.54. Since the line is matched at the source, the voltage sent out initially is $\frac{1}{2}V_0$. The open-circuited load has a voltage reflection coefficient of $\Gamma_L = +1$. Hence, at the load, the incident wave, $\frac{1}{2}V_0$, is reflected completely, giving a total voltage at the load of V_0 as desired, as shown in Fig. 2.55. Since the source is matched at the source, there will be

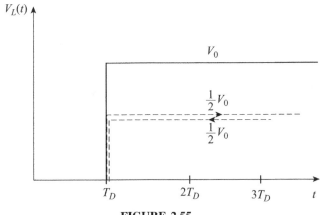

FIGURE 2.55

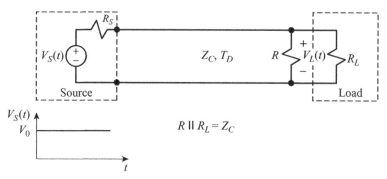

FIGURE 2.56. Parallel matching at the load end.

no reflections there, so that the load voltage will be brought up to the desired voltage of V_0.

Parallel Matching at the Load End (Fig. 2.56) The voltage sent out initially is

$$V_{\text{init}} = \frac{Z_C}{R_S + Z_C} V_0$$

Since the line is matched at the load, there are no reflections there, as shown in Fig. 2.57.

There are two disadvantages to the parallel match scheme compared to the series match scheme. When the source and load are CMOS, their input

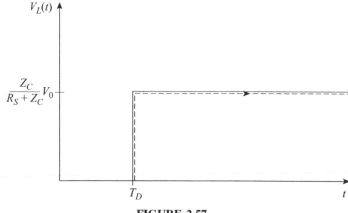

FIGURE 2.57

impedances look capacitive and in the steady state appear as an open circuit and do not draw current.

1. When the load is a CMOS gate, the R at the load draws current and dissipates unnecessary power.
2. The steady-state voltage at the load is

$$V_{L,ss} = \frac{(R\|R_L = Z_C)}{(R\|R_L = Z_C) + R_S} V_0$$

which is less than the desired V_0 that the series match achieves.

Examples of Signal Integrity Matching   Figure 2.58 shows a line that is mismatched at the source and the load. The PSPICE program is

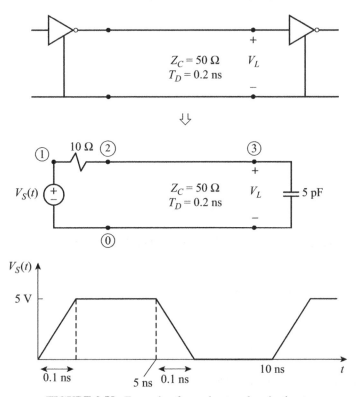

FIGURE 2.58. Example of overshoot and undershoot.

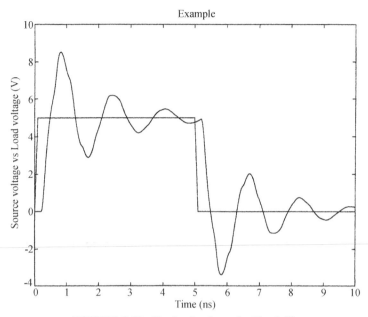

FIGURE 2.59. The load voltage for Fig. 2.58.

```
EXAMPLE
VS 1 0 PULSE(0 5 0 0.1N 0.1N 4.9N 10N)
RS 1 2 10
T 2 0 3 0 Z0=50 TD=0.2N
CL 3 0 5P
.TRAN 0.01N 10N 0 0.01N
.PROBE
.END
```

The load voltage is shown in Fig. 2.59.

Series Match at the Source(Fig. 2.60) The PSPICE program is

```
EXAMPLE
VS 1 0 PULSE(0 5 0 0.1N 0.1N 4.9N 10N)
RS 1 2 10
R 2 3 40
T 3 0 4 0 Z0=50 TD=0.2N
CL 4 0 5P
.TRAN 0.01N 10N 0 0.01N
.PROBE
.END
```

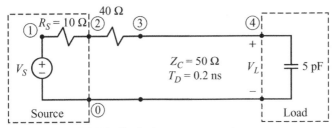

FIGURE 2.60. Series matching at the source.

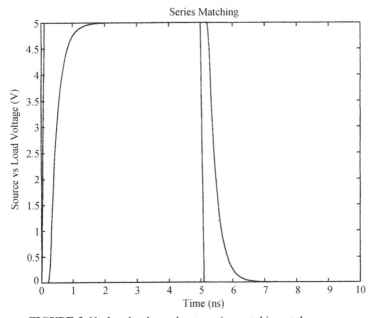

FIGURE 2.61. Load voltage due to series matching at the source.

The load voltage is shown in Fig. 2.61. Since the source is matched, the load capacitance "sees" a Thévenin impedance of Z_C and therefore its voltage increases with a time constant of $Z_C C_L = 0.25$ ns.

Parallel Match at the Load (Fig. 2.62)   The PSPICE program is

```
EXAMPLE
VS 1 0 PULSE(0 5 0 0.1N 0.1N 4.9N 10N)
RS 1 2 10
T 2 0 3 0 Z0=50 TD=0.2N
CL 3 0 5P
R 3 0 50
.TRAN 0.01N 10N 0 0.01N
.PROBE
.END
```

The load voltage is shown in Fig. 2.63.

What causes the "blip" in the load voltage? Isn't the line matched at the load? *Note:* The impedance of the capacitor is smaller than that of the resistor above the following frequency, and hence the line is not matched at these frequency components of $V_S(t)$ above:

$$f = \frac{1}{2\pi RC} = \frac{1}{2\pi \times 50 \times 5 \times 10^{-12}} = 637\,\text{MHz}$$

But since the source is not matched for these higher-frequency components of $V_S(t)$, these reflections at the load are re-reflected at the source and keep

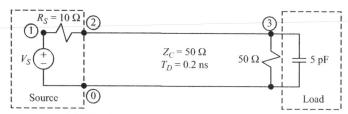

FIGURE 2.62. Parallel matching at the load.

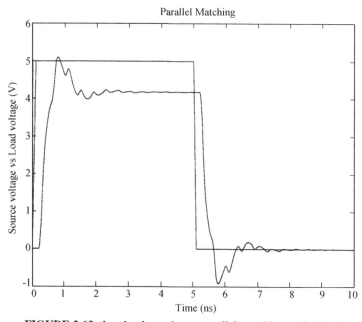

FIGURE 2.63. Load voltage due to parallel matching at the load.

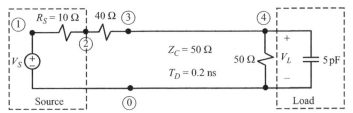

FIGURE 2.64. Matching at both ends.

bouncing back and forth along the line and cause the "blip." The bandwidth of $V_S(t)$ is $\mathrm{BW} = 1/\tau_r = 10\,\mathrm{GHz}$, so there are a significant number of harmonics that are not matched by this scheme. The steady-state load voltage is 4.167 V.

Series Matching at the Source End and Parallel Matching at the Load End (Fig. 2.64)   Since matching at the source and the load works well, one might wonder if matching at both ends as shown in Fig. 2.64 might achieve both objectives. This is not the case, as the following example shows. Since the source is matched, the voltage sent out initially is $V_0/2$, and since the load is not matched at certain high frequencies, there is no reflection to bring that up to the desired V_0. Since the source is matched, these reflections at the load are absorbed at the source and cease being reflected at the source. The load voltage is $V_0/2 = 2.5$ V, as shown in Fig. 2.65, and logic errors are likely to result!

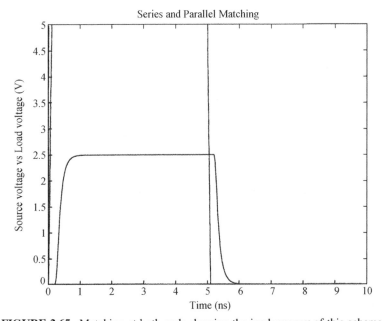

FIGURE 2.65. Matching at both ends showing the inadequency of this scheme.

2.10   EFFECT OF LINE DISCONTINUITIES

Line discontinuities occur in cables in the form of connectors and on PCBs in the form of abrupt changes in land widths as well as the presence of vias between layers in innerlayer PCBs. In this section we will see that line discontinuities also cause reflections and signal integrity problems in the same way that terminal discontiuities do.

So far we have been considering *uniform lines* where the line cross section is uniform along its length; that is, its properties (radius of wires, separation of wires, land widths, and land separations) do not change along its length. Hence its per-unit-length parameters of inductance l and capacitance c do not depend on position z along its length. This greatly simplifies the solution of the transmission-line equations that contain them.

Suppose that the line changes its Z_C at some point along its length, perhaps by abruptly widening out, as illustrated in Fig. 2.66. At the discontinuity, a wave coming from the left essentially sees the right half as a resistance Z_{C2} terminating the left half. So there is a voltage reflection coefficient for voltage waves incident from the left side of

$$\boxed{\Gamma_{12} = \frac{Z_{C2} - Z_{C1}}{Z_{C2} + Z_{C1}}} \tag{2.85}$$

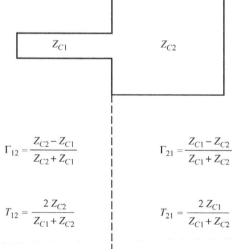

$$\Gamma_{12} = \frac{Z_{C2} - Z_{C1}}{Z_{C2} + Z_{C1}} \qquad\qquad \Gamma_{21} = \frac{Z_{C1} - Z_{C2}}{Z_{C1} + Z_{C2}}$$

$$T_{12} = \frac{2 Z_{C2}}{Z_{C1} + Z_{C2}} \qquad\qquad T_{21} = \frac{2 Z_{C1}}{Z_{C1} + Z_{C2}}$$

FIGURE 2.66. Changes in characteristic impedance by an abrupt change in land width on a PCB.

Similarly, for voltage waves traveling to the left in the second line, there is a voltage reflection coefficient on the right-hand side of the discontinuity of

$$\Gamma_{21} = \frac{Z_{C1} - Z_{C2}}{Z_{C1} + Z_{C2}} \qquad (2.86)$$

In addition to the voltages reflected on each side of the discontinuity, there are also voltage waves *transmitted across the discontinuity*. This is shown in Fig. 2.67.

The *sum* of the incident and reflected voltages on one side of the discontinuity must equal the transmitted voltage on the other side of the discontinuity. For voltage waves incident from the left side, we must have

$$v_{i1} + v_{r1} = v_{t2}$$
$$1 + \frac{v_{r1}}{v_{i1}} = \frac{v_{t2}}{v_{i1}} \qquad (2.87)$$

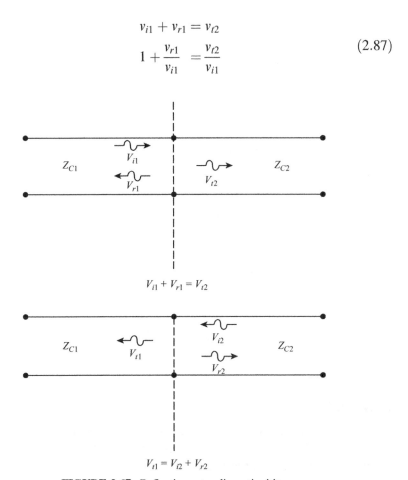

FIGURE 2.67. Reflections at a discontinuitity.

or

$$\boxed{1 + \Gamma_{12} = T_{12}}$$ (2.88)

Hence we obtain the *voltage transmission coefficients* on either side of the boundary (of exactly the same form as the transmission coefficients for uniform plane waves incident normal to a boundary):

$$\boxed{T_{12} = \frac{2Z_{C2}}{Z_{C2} + Z_{C1}}}$$ (2.89)

Similarly, for waves traveling to the left and incident on the discontinuity from the right side, we must have

$$v_{i2} + v_{r2} = v_{t1}$$ (2.90)

So we obtain the voltage transmission coefficient

$$\boxed{T_{21} = \frac{2Z_{C1}}{Z_{C1} + Z_{C2}}}$$ (2.91)

and

$$\boxed{1 + \Gamma_{21} = T_{21}}$$ (2.92)

EXAMPLE

Sketch the line voltages at the midpoint and endpoints of the line in Fig. 2.68.

We simply sketch the waves on the line at points AA' and BB' and then add all the waves present on the graph.

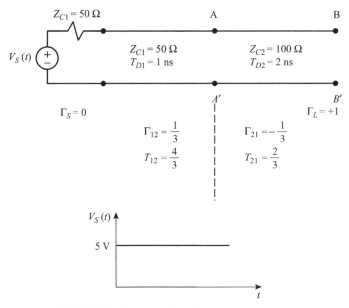

FIGURE 2.68. Example of a line discontinuity.

The voltage sent out initially is

$$V_{\text{init}} = \frac{50}{50 + 50} \, V_S(t)$$

$$= 2.5 \text{ V}$$

The graphs are given in Fig. 2.69.
  The PSPICE program is

```
EXAMPLE
VS 1 0 PWL(0 0 1P 5 100N 5)
RS 1 2 50
T 2 0 3 0 Z0=50 TD=1N
T 3 0 4 0 Z0=100 TD=2N
RL 4 0 1E8
.TRAN 0.1N 30N 0 0.1N
.PROBE
.END
```

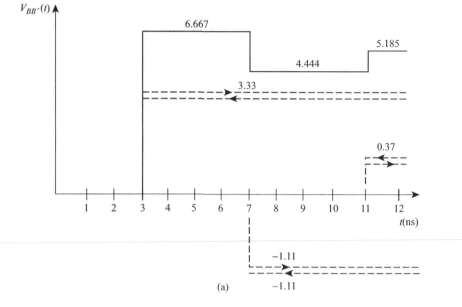

(a)

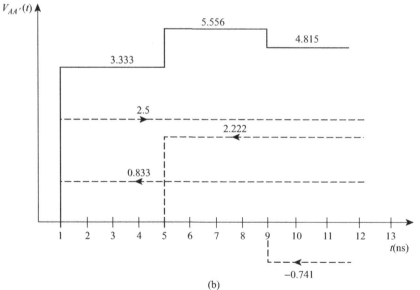

(b)

FIGURE 2.69. Line voltages for the discontinuity in Fig. 2.68.

The line voltages at the discontinuities so obtained using PSPICE are sketched in Fig. 2.70.

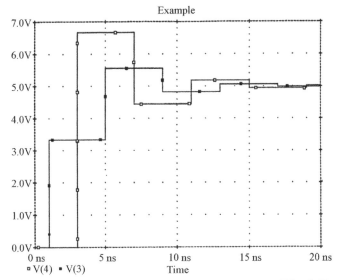

FIGURE 2.70. PSPICE solution for the line voltages of Fig. 2.68.

2.11   DRIVING MULTIPLE LINES

It is very common to have one source, such as a digital clock, driving numerous loads that are either in series or in parallel. Analysis of these by hand as we have done before is very involved, and little design information can be obtained from this analysis. The simplest way to analyze these configurations and determine their performance is to model *each section* with PSPICE, interconnect these models in one PSPICE program, and then run the entire program to view the intermediate results to determine the resulting voltages and currents, as we did in the preceding example.

3

FREQUENCY-DOMAIN ANALYSIS OF TWO-CONDUCTOR LINES

In this chapter we examine the *frequency response* of a two-conductor transmission line to a single-frequency sinusoidal source: $V_S(t) = V_S \cos(\omega t + \theta_S)$. The time-domain line voltages and currents, $V(z, t)$ and $I(z, t)$, at each point along the line, z (generally of interest only at the line terminals), are to be determined and are similarly single-frequency sinusoids at the same frequency as the source and have a magnitude and a phase angle that are functions of their position along the line, z.

To simplify the solutions for the line voltage and current, we examine the *steady-state phasor* response of the line, as illustrated in Fig. 3.1. We replace the time-domain source with the complex, phasor source (denoted with a caret) as

$$V_S(t) = V_S \cos(\omega t + \theta_S) \Rightarrow \underbrace{V_S \angle \theta_S}_{\hat{V}_S} e^{j\omega t} \tag{3.1}$$

Similarly, the line voltages and currents are replaced with their *phasor* forms:

$$V(z, t) = V(z)\cos(\omega t + \theta_V(z)) \Rightarrow \underbrace{V \angle \theta_V}_{\hat{V}(z)} e^{j\omega t} \tag{3.2a}$$

Transmission Lines in Digital Systems for EMC Practitioners, First Edition.
Clayton R. Paul.
© 2012 John Wiley & Sons, Inc. Published 2012 by John Wiley & Sons, Inc.

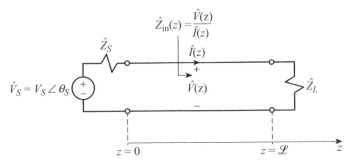

FIGURE 3.1. Sinusoidal steady-state (phasor) solution.

$$I(z, t) = I(z)\cos(\omega t + \theta_I(z)) \Rightarrow \underbrace{I \angle \theta_I}_{\hat{I}(z)} e^{j\omega t} \tag{3.2b}$$

3.1   THE TRANSMISSION-LINE EQUATIONS FOR SINUSOIDAL STEADY-STATE (PHASOR) EXCITATION OF THE LINE

The time-domain derivatives are replaced with $j\omega$:

$$\frac{\partial}{\partial t} \Rightarrow j\omega \tag{3.3}$$

where the radian frequency of the source is $\omega = 2\pi f$ rad/s and f is its cyclic frequency in hertz. The (phasor) transmission-line equations for a *lossless* line become

$$\boxed{\frac{d\hat{V}(z)}{dz} = -j\omega l \hat{I}(z)} \tag{3.4a}$$

$$\boxed{\frac{d\hat{I}(z)}{dz} = -j\omega c \hat{V}(z)} \tag{3.4b}$$

The uncoupled second-order transmission-line equations become

$$\boxed{\frac{d^2\hat{V}(z)}{dz^2} + \underbrace{\omega^2 lc}_{\beta^2} \hat{V}(z) = 0} \tag{3.5a}$$

$$\frac{d^2 \hat{I}(z)}{dz^2} + \underbrace{\omega^2 lc}_{\beta^2} \hat{I}(z) = 0 \qquad (3.5b)$$

3.2   THE GENERAL SOLUTION FOR THE LINE VOLTAGES AND CURRENTS

The general phasor solutions to these phasor equations are easily obtained as

$$\hat{V}(z) = \hat{V}^+ e^{-j\beta z} + \hat{V}^- e^{j\beta z} \qquad (3.6a)$$

$$\hat{I}(z) = \frac{\hat{V}^+}{Z_C} e^{-j\beta z} - \frac{\hat{V}^-}{Z_C} e^{j\beta z} \qquad (3.6b)$$

where the *phase constant* is

$$\begin{aligned} \beta &= \omega\sqrt{lc} \\ &= \frac{\omega}{v} \quad \text{rad/m} \end{aligned} \qquad (3.7)$$

The time-domain solutions are

$$V(z,t) = V^+ \cos(\omega t - \beta z + \theta^+) + V^- \cos(\omega t + \beta z + \theta^-) \qquad (3.8a)$$

$$I(z,t) = \frac{V^+}{Z_C} \cos(\omega t - \beta z + \theta^+) - \frac{V^-}{Z_C} \cos(\omega t + \beta z + \theta^-) \qquad (3.8b)$$

and the undetermined constants in the general solution, $\hat{V}^+ = V^+ \angle \theta^+$ and $\hat{V}^- = V^- \angle \theta^-$, have a magnitude and a phase.

3.3   THE VOLTAGE REFLECTION COEFFICIENT AND INPUT IMPEDANCE OF THE LINE

The voltage reflection coefficient at a point z on the line is defined as the ratio of the reflected wave, $\hat{V}^- e^{j\beta z}$, and the incident wave, $\hat{V}^+ e^{-j\beta z}$, at that point:

$$
\begin{aligned}
\hat{\Gamma}(z) &= \frac{\hat{V}^- e^{j\beta z}}{\hat{V}^+ e^{-j\beta z}} \\
&= \frac{\hat{V}^-}{\hat{V}^+} e^{j2\beta z}
\end{aligned}
\tag{3.9}
$$

This is a general voltage reflection coefficient at any z along the line. Evaluating this at the load, $z = \mathscr{L}$, gives

$$
\hat{\Gamma}_L = \frac{\hat{V}^-}{\hat{V}^+} e^{j2\beta \mathscr{L}}
\tag{3.10}
$$

where the voltage reflection coefficient at the load is, as before,

$$
\boxed{\hat{\Gamma}_L = \frac{\hat{Z}_L - Z_C}{\hat{Z}_L + Z_C}}
\tag{3.11}
$$

The voltage reflection coefficient at any point along the line can then be written in terms of the load reflection coefficient as

$$
\boxed{\hat{\Gamma}(z) = \hat{\Gamma}_L e^{j2\beta\,(z-\mathscr{L})}}
\tag{3.12}
$$

The general solutions at any point on the line can be written in terms of the voltage reflection coefficient there as

$$
\boxed{\hat{V}(z) = \hat{V}^+ e^{-j\beta z}[1 + \hat{\Gamma}(z)]}
\tag{3.13a}
$$

$$\hat{I}(z) = \frac{\hat{V}^+}{Z_C} e^{-j\beta z} \left[1 - \hat{\Gamma}(z) \right] \qquad (3.13b)$$

Substituting the explicit relation for the reflection coefficient in terms of the load reflection given in (3.12) gives

$$\hat{V}(z) = \hat{V}^+ e^{-j\beta z} \left[1 + \hat{\Gamma}_L e^{j2\beta(z-\mathscr{L})} \right] \qquad (3.14a)$$

$$\hat{I}(z) = \frac{\hat{V}^+}{Z_C} e^{-j\beta z} \left[1 - \hat{\Gamma}_L e^{j2\beta(z-\mathscr{L})} \right] \qquad (3.14b)$$

The input impedance at any point on the line is defined as

$$
\begin{aligned}
\hat{Z}_{in}(z) &= \frac{\hat{V}(z)}{\hat{I}(z)} \\
&= Z_C \frac{1 + \hat{\Gamma}(z)}{1 - \hat{\Gamma}(z)}
\end{aligned} \qquad (3.15)
$$

The reflection coefficients at the input and at the load are related by

$$\hat{\Gamma}(0) = \hat{\Gamma}_L e^{-j2\beta\mathscr{L}} \qquad (3.16)$$

Hence the input impedance to a line of length $\mathscr{L}$ is

$$
\begin{aligned}
\hat{Z}_{in} &= Z_C \frac{1 + \hat{\Gamma}(0)}{1 - \hat{\Gamma}(0)} \\
&= Z_C \frac{1 + \hat{\Gamma}_L e^{-j2\beta\mathscr{L}}}{1 - \hat{\Gamma}_L e^{-j2\beta\mathscr{L}}}
\end{aligned} \qquad (3.17)
$$

where $\hat{Z}_{in} \equiv \hat{Z}_{in}(0)$.

3.4   THE SOLUTION FOR THE TERMINAL VOLTAGES AND CURRENTS

The phasor solutions for the terminal voltages and currents are

$$\hat{V}(0) = \hat{V}^{+}[1 + \hat{\Gamma}(0)] \tag{3.18a}$$

$$\hat{I}(0) = \frac{\hat{V}^{+}}{Z_C}[1 - \hat{\Gamma}(0)] \tag{3.18b}$$

and

$$\hat{V}(\mathscr{L}) = \hat{V}^{+}e^{-j\beta\mathscr{L}}(1 + \hat{\Gamma}_L) \tag{3.19a}$$

$$\hat{I}(\mathscr{L}) = \frac{\hat{V}^{+}}{Z_C}e^{-j\beta\mathscr{L}}(1 - \hat{\Gamma}_L) \tag{3.19b}$$

We only need to determine $\hat{V}^{+}$. The input to the phasor line appears as shown in Fig. 3.2. Hence, by voltage division we can determine the phasor input voltage to the line as

$$\hat{V}(0) = \frac{\hat{Z}_{in}}{\hat{Z}_S + \hat{Z}_{in}}\hat{V}_S \tag{3.20}$$

and the undetermined constant is determined from (3.18a) as

$$\hat{V}^{+} = \frac{\hat{V}(0)}{1 + \hat{\Gamma}(0)} \tag{3.21}$$

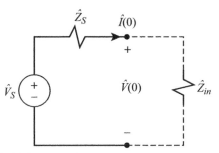

FIGURE 3.2. Equivalent input circuit to the phasor line.

EXAMPLE

Consider the circuit shown in Fig. 3.3. How long is the line in wavelengths?

$$\mathscr{L} = k\frac{v}{f}$$

$$= 1.35\lambda$$

The load voltage reflection coefficient is

$$\hat{\Gamma}_L = \frac{\hat{Z}_L - Z_C}{\hat{Z}_L + Z_C}$$

$$= \frac{50 + j200}{150 + j200}$$

$$= 0.82 \angle 22.83°$$

and the input voltage reflection coefficient is

$$\hat{\Gamma}(0) = \hat{\Gamma}_L e^{-j2\beta\mathscr{L}}$$

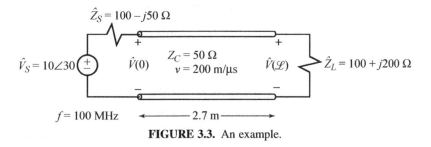

FIGURE 3.3. An example.

But

$$-2\beta \mathscr{L} = -4\pi f \frac{\mathscr{L}}{v}$$
$$= -4\pi \frac{\mathscr{L}}{\lambda}$$
$$= -5.4\pi \qquad \text{rad}$$
$$= -972°$$

Therefore,

$$\hat{\Gamma}(0) = \hat{\Gamma}_L e^{-j2\beta \mathscr{L}}$$
$$= 0.82 \angle - 949.17°$$

and the input impedance to the line is

$$\hat{Z}_{in} = Z_C \frac{1 + \hat{\Gamma}(0)}{1 - \hat{\Gamma}(0)}$$
$$= 23.35 \angle 75.62°$$

Therefore, the input voltage to the line is

$$\hat{V}(0) = \frac{\hat{Z}_{in}}{\hat{Z}_S + \hat{Z}_{in}} \hat{V}_S$$
$$= 2.14 \angle 120.13°$$

The undetermined constant is

$$\hat{V}^+ = \frac{\hat{V}(0)}{1 + \hat{\Gamma}(0)}$$
$$= 2.75 \angle 66.58°$$

The input and load phasor voltages are

$$\hat{V}(0) = \hat{V}^+[1 + \hat{\Gamma}(0)]$$
$$= 2.14 \angle 120.13°$$
$$\hat{V}(\mathscr{L}) = \hat{V}^+ e^{-j\beta \mathscr{L}}(1 + \hat{\Gamma}_L)$$
$$= 4.93 \angle - 409.12°$$

Hence the time-domain terminal voltages are

$$V(0, t) = 2.14\cos(6.28 \times 10^8 t + 120.13°)$$
$$V(\mathscr{L}, t) = 4.93\cos(6.28 \times 10^8 t - 49.12°)$$

3.5   THE SPICE SOLUTION

There are three changes from the time-domain use of PSPICE for the phasor use of PSPICE.

1. The specification of the voltage source is changed to

VS   N1   N2   AC   mag   phase

2. The .TRAN line needs to be changed to

.AC   DEC   1   f   f

where f is the frequency of the source.

3. The output is obtained as

.PRINT   AC   VM(NX)   VP(NX)

where VM is the magnitude of the phasor voltage at node NX and VP is the phase.

EXAMPLE

To solve the previous phasor example using PSPICE, we must generate equivalent circuits to represent the complex impedances $\hat{Z}_S$ and $\hat{Z}_L$ using lumped-circuit elements such as R's, L's, and C's (Fig. 3.4) that give the same complex impedances at the source frequency.

The PSPICE program is

```
EXAMPLE
VS 1 0 AC 10 30
RS 1 2 100
CS 2 3 31.8P
T 3 0 4 0 Z0=50 TD=13.5N
RL 4 5 100
LL 5 0 0.318U
.AC DEC 1 1E8 1E8.
PRINT AC VM(3) VP(3) VM(4) VP(4)
+ IM(CS) IP(CS)
.END
```

giving

$$V(3) = 2.136 \angle 120.1°, \quad V(4) = 4.926 \angle -49.1°$$

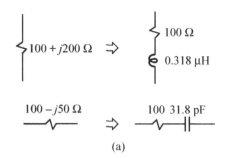

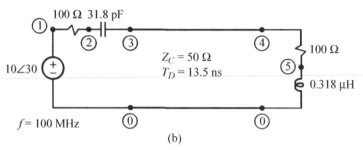

FIGURE 3.4. Lumped equivalent circuit for the phasor circuit of Fig. 3.3.

and the input impedance to the line is computed as

$$\hat{Z}_{in} = \frac{VM(3)}{IM(CS)} \angle VP(3) - IP(CS) = 23.349 \angle 75.56°$$

as computed by hand.

3.6 VOLTAGE AND CURRENT AS A FUNCTION OF POSITION ON THE LINE

Previously we determined the voltage and current for sinusoidal excitation only at the endpoints of the line. *How do they vary along the line?* The voltage and current along the line are

$$\hat{V}(z) = \hat{V}^+ e^{-j\beta z}\left[1 + \hat{\Gamma}_L e^{j2\beta(z-\mathscr{L})}\right] \tag{3.22a}$$

$$\hat{I}(z) = \frac{\hat{V}^+}{Z_C} e^{-j\beta z}\left[1 - \hat{\Gamma}_L e^{j2\beta(z-\mathscr{L})}\right] \tag{3.22b}$$

We will plot the magnitude of these for distances $d = \mathcal{L} - z$ away from the load. Taking the magnitudes of the phasor voltage and current gives

$$\left|\hat{V}(d = \mathcal{L} - z)\right| = \left|\hat{V}^+\right|\left|1 + \hat{\Gamma}_L e^{-j2\beta d}\right| \tag{3.23a}$$

$$\left|\hat{I}(d = \mathcal{L} - z)\right| = \frac{\left|\hat{V}^+\right|}{Z_C}\left|1 - \hat{\Gamma}_L e^{-j2\beta d}\right| \tag{3.23b}$$

There are three important cases of special interest that we will investigate, shown in Fig. 3.5: (1) the load is a short circuit, $\hat{Z}_L = 0$; (2) the load is an open circuit, $\hat{Z}_L = \infty$; and (3) the load is matched, $\hat{Z}_L = Z_C$.

For the case where the load is a short circuit, $\hat{Z}_L = 0$, the load reflection coefficient is $\hat{\Gamma}_L = -1$ and the equations reduce to

$$
\begin{aligned}
\left|\hat{V}(d)\right| &= \left|\hat{V}^+\right|\left|1 - e^{-j2\beta d}\right| \\
&= \left|\hat{V}^+\right|\underbrace{\left|e^{-j\beta d}\right|}_{1}\underbrace{\left|e^{j\beta d} - e^{-j\beta d}\right|}_{|2j\sin(\beta d)|} \\
&\propto \left|\sin(\beta d)\right| \\
&= \left|\sin\left(2\pi\frac{d}{\lambda}\right)\right|
\end{aligned}
$$

and

$$
\begin{aligned}
\left|\hat{I}(d)\right| &= \frac{\left|\hat{V}^+\right|}{Z_C}\left|1 + e^{-j2\beta d}\right| \\
&= \frac{\left|\hat{V}^+\right|}{Z_C}\underbrace{\left|e^{-j\beta d}\right|}_{1}\underbrace{\left|e^{j\beta d} + e^{-j\beta d}\right|}_{|2\cos(\beta d)|} \\
&\propto \left|\cos(\beta d)\right| \\
&= \left|\cos\left(2\pi\frac{d}{\lambda}\right)\right|
\end{aligned}
$$

The voltage is zero at points away from the short circuit load, where d is a multiple of a half-wavelength and a maximum at points where d is an odd multiple of a quarter-wavelength.

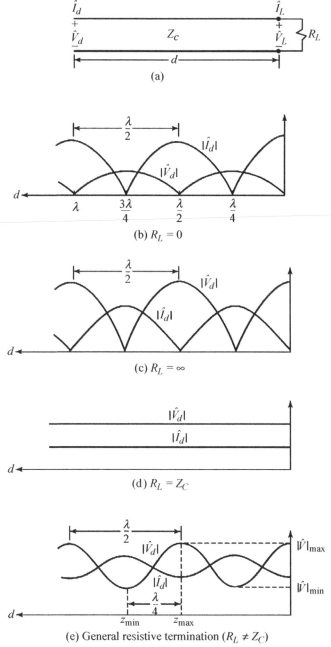

FIGURE 3.5. Voltage and current at various points on the line.

For the case where the load is an open circuit, $\hat{Z}_L = \infty$, the load reflection coefficient is $\hat{\Gamma}_L = 1$ and the equations reduce to

$$\left|\hat{V}(d)\right| = \left|\hat{V}^+\right|\left|1 + e^{-j2\beta d}\right|$$

$$\propto \left|\cos(\beta d)\right|$$

$$= \left|\cos\left(2\pi\frac{d}{\lambda}\right)\right|$$

and

$$\left|\hat{I}(d)\right| = \frac{\left|\hat{V}^+\right|}{Z_C}\left|1 - e^{-j2\beta d}\right|$$

$$\propto \left|\sin(\beta d)\right|$$

$$= \left|\sin\left(2\pi\frac{d}{\lambda}\right)\right|$$

The current is zero at points away from the short circuit load where d is multiple of a half-wavelength and a maximum at points where d is an odd multiple of a quarter-wavelength.

For a matched line there is no variation in the voltage and current along the line. That's the advantage of being *matched*, $R_L = Z_C$.

Notice that whatever the value of the load impedance, the voltage (current) maximum and adjacent voltage (current) minimum are separated by $\lambda/4$. The corresponding voltage or current points are separated by $\lambda/2$.

From (3.17), the input impedance to the line replicates for line lengths that are a multiple of $\lambda/2$ since $2\beta\mathscr{L} = 4\pi(\mathscr{L}/\lambda) = 2\pi$ and is shown in Fig. 3.6.

3.7   MATCHING AND VSWR

If the line is not matched, how do we *quantitatively* judge the *degree of the mismatch*? The answer is the voltage standing-wave ratio (VSWR), defined as

$$\text{VSWR} = \frac{|\hat{V}|_{\text{max}}}{|\hat{V}|_{\text{min}}} \tag{3.24}$$

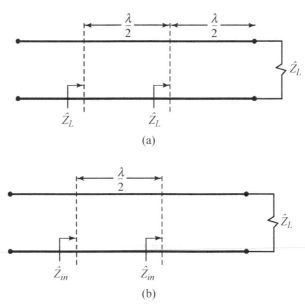

(a)

(b)

FIGURE 3.6. Replication of the input impedance for line lengths that are a multiple of a half-wavelength.

A maximum and the adjacent minimum are separated by exactly $\lambda/4$.

For a short-circuit load, $\hat{Z}_L = 0$, the voltage minimum is zero, so the VSWR is infinite. Also for an open-circuit load, $\hat{Z}_L = \infty$, the voltage minimum is zero, so the VSWR is infinite. So

$$\text{VSWR} = \begin{cases} \infty & \hat{Z}_L = 0 \\ \infty & \hat{Z}_L = \infty \end{cases}$$

but the VSWR is unity for a matched load:

$$\text{VSWR} = 1 \qquad \hat{Z}_L = Z_C$$

The VSWR must therefore lie between these two bounds:

$$1 \leq \text{VSWR} < \infty \tag{3.25}$$

Industry considers a line to be "matched" if VSWR < 1.2. The VSWR can be calculated directly by taking the ratios of the *magnitudes* of the maximum and minimum of (3.14a):

$$\boxed{\text{VSWR} = \frac{1 + |\hat{\Gamma}_L|}{1 - |\hat{\Gamma}_L|}} \tag{3.26}$$

3.8 POWER FLOW ON THE LINE

At any point on the line, the power flow to the right as illustrated in Fig. 3.7 is

$$P_{AV}(z) = \frac{1}{2} \text{Re}\left\{\hat{V}(z)\hat{I}^*(z)\right\} \tag{3.27}$$

where * denotes the complex conjugate. Hence

$$
\begin{aligned}
P_{AV}(z) &= \frac{1}{2} \text{Re}\left\{\hat{V}(z)\hat{I}^*(z)\right\} \\
&= \frac{1}{2} \text{Re}\left\{\hat{V}^+ e^{-j\beta z}\left[1 + \hat{\Gamma}_L e^{j2\beta(z-\mathscr{L})}\right]\frac{\hat{V}^{+^*}}{Z_C}e^{j\beta z}\left[1 - \hat{\Gamma}_L^* e^{-j2\beta(z-\mathscr{L})}\right]\right\} \\
&= \frac{1}{2} \text{Re}\left\{\frac{\hat{V}^+ \hat{V}^{+^*}}{Z_C}\left[1 + \underbrace{\left(\hat{\Gamma}_L e^{j2\beta(z-\mathscr{L})} - \hat{\Gamma}_L^* e^{-j2\beta(z-\mathscr{L})}\right)}_{\text{imaginary}} - \hat{\Gamma}_L \hat{\Gamma}_L^*\right]\right\} \\
&= \frac{|\hat{V}^+|^2}{2Z_C}\left(1 - |\hat{\Gamma}_L|^2\right)
\end{aligned}
$$

$$\tag{3.28}$$

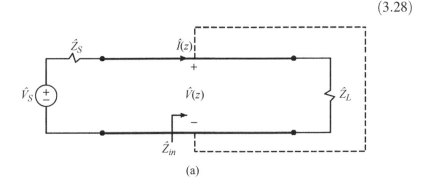

(a)

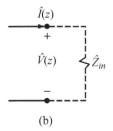

(b)

FIGURE 3.7. Power flow to the right on the line.

But the power in the forward-traveling wave is

$$P_{AV}^+(z) = \frac{1}{2} \mathrm{Re}\left\{ \hat{V}^+ e^{-j\beta z} \left(\frac{\hat{V}^{+^*}}{Z_C} e^{j\beta z} \right) \right\}$$

$$= \frac{|\hat{V}^+|^2}{2Z_C} \tag{3.29}$$

and the power in the backward-traveling wave is

$$P_{AV}^-(z) = \frac{1}{2} \mathrm{Re}\left\{ \hat{V}^- e^{j\beta z} \left(-\frac{\hat{V}^{-^*}}{Z_C} e^{-j\beta z} \right) \right\}$$

$$= -\frac{|\hat{V}^+|^2}{2Z_C} |\hat{\Gamma}_L|^2 \tag{3.30}$$

So the portion of the incident power that is reflected at the load is $|\hat{\Gamma}_L|^2 \times 100\%$.

If the line is *lossless*, the total power delivered to the load is equal to the total power delivered to the input to the line, as shown in Fig. 3.8!

So for a *lossless* line, to determine the power delivered to the load you simply need to determine the power delivered to the input to the line:

$$P_{AV,\ \text{to load}} = P_{AV,\ \text{to line input}} = \frac{1}{2} \frac{|\hat{V}(0)|^2}{|\hat{Z}_{in}|} \cos(\theta_{Z_{in}}) \tag{3.31}$$

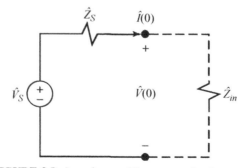

FIGURE 3.8. Impedance of line as seen by the source.

EXAMPLE

In the preceding example we obtained

$$\hat{V}(0) = \hat{V}^+[1 + \hat{\Gamma}(0)]$$
$$= 2.14 \angle 120.13°$$
$$\hat{V}(\mathscr{L}) = \hat{V}^+ e^{-j\beta\mathscr{L}}(1 + \hat{\Gamma}_L)$$
$$= 4.93 \angle -409.12°$$
$$\hat{Z}_{in} = Z_C \frac{1 + \hat{\Gamma}(0)}{1 - \hat{\Gamma}(0)}$$
$$= 23.35 \angle 75.62°$$

Hence the power delivered to the line input and the load (since the line is lossless) is

$$P_{AV, \text{ to input}} = \frac{1}{2} \frac{|\hat{V}(0)|^2}{|\hat{Z}_{in}|} \cos(\theta_{Z_{in}})$$

$$= \frac{1}{2} \frac{(2.14)^2}{23.35} \cos(75.62°)$$

$$= 24.3 \text{ mW}$$

and (directly)

$$P_{AV, \text{ to load}} = \frac{1}{2} \frac{|\hat{V}(\mathscr{L})|^2}{|\hat{Z}_L|} \cos(\theta_{Z_L})$$

$$= \frac{1}{2} \frac{(4.93)^2}{223.6} \cos(63.43°)$$

$$= 24.3 \text{ mW}$$

where $\hat{Z}_L = 100 + j200 = 223.6 \angle 63.43°$. The VSWR is ($\hat{\Gamma}_L = 0.82 \angle 22.83°$)

$$\text{VSWR} = \frac{1 + |\hat{\Gamma}_L|}{1 - |\hat{\Gamma}_L|}$$

$$= \frac{1 + 0.82}{1 - 0.82}$$

$$= 10.1$$

3.9   ALTERNATIVE FORMS OF THE RESULTS

The previous results for the line voltages and currents and the input impedance are the simplest to use. However, alternative formulas can be obtained (you should derive these):

$$\hat{V}(z) = \frac{1 + \hat{\Gamma}_L e^{-j2\beta\mathscr{L}} e^{j2\beta z}}{1 - \hat{\Gamma}_S \hat{\Gamma}_L e^{-j2\beta\mathscr{L}}} \frac{Z_C}{\hat{Z}_S + Z_C} \hat{V}_S e^{-j\beta z} \tag{3.32}$$

$$\hat{I}(z) = \frac{1 - \hat{\Gamma}_L e^{-j2\beta\mathscr{L}} e^{j2\beta z}}{1 - \hat{\Gamma}_S \hat{\Gamma}_L e^{-j2\beta\mathscr{L}}} \frac{1}{\hat{Z}_S + Z_C} \hat{V}_S e^{-j\beta z} \tag{3.33}$$

$$\hat{Z}_{in} = Z_C \frac{\hat{Z}_L + j Z_C \tan(\beta\mathscr{L})}{Z_C + j\hat{Z}_L \tan(\beta\mathscr{L})} \tag{3.34}$$

These give considerable insight into the line behavior. You should check these for matched lines, $\hat{Z}_L = Z_C$.

3.10   CONSTRUCTION OF MICROWAVE CIRCUIT COMPONENTS USING TRANSMISSION LINES

The inductance and capacitance of the leads attaching a discrete resistor, capacitor, or inductor to the circuit can cause significant deterioration of its performance at microwave frequencies. Shorted transmission lines on PCBs are used to simulate lumped-circuit elements in microwave circuits.

The input impedance to a transmission line in its most general form is

$$\hat{Z}_{in} = Z_C \frac{1 + \hat{\Gamma}_L e^{-j2\beta\mathscr{L}}}{1 - \hat{\Gamma}_L e^{-j2\beta\mathscr{L}}}$$

Consider the case of a line that is a quarter-wavelength long, $\mathscr{L} = \lambda/4$, shown in Fig. 3.9. In the case of a short-circuit load,

$$\hat{\Gamma}_L = \frac{0 - Z_C}{0 + Z_C}$$

$$= -1$$

$$e^{-j2\beta\mathscr{L}} = e^{-j4\pi(\mathscr{L}/\lambda)}$$

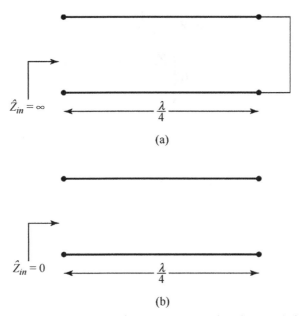

FIGURE 3.9. Impedances seen into a quarter-wavelength transmission line.

$$e^{-j2\beta\mathscr{L}} = e^{-j\pi}$$
$$= -1 \qquad \mathscr{L} = \frac{1}{4}\lambda$$

$$\hat{\Gamma}_{in} = \hat{\Gamma}_L e^{-j2\beta\mathscr{L}}$$
$$= (-1)(-1)$$
$$= 1$$

$$\hat{Z}_{in} = Z_C \frac{1+1}{1-1}$$
$$= \infty \begin{cases} \hat{Z}_L = \text{short circuit} \\ \mathscr{L} = \dfrac{\lambda}{4} \end{cases}$$

Hence a quarter-wavelength line having a short-circuit load appears as a open circuit!

In the case of an open-circuit load, $\Gamma_L = +1$:

$$\hat{\Gamma}_{in} = \hat{\Gamma}_L e^{-j2\beta\mathscr{L}}$$
$$= (+1)(-1)$$
$$= -1$$

$$\hat{Z}_{in} = Z_C \frac{1-1}{1+1}$$

$$= 0 \begin{cases} \hat{Z}_L = \text{open circuit} \\ \mathscr{L} = \dfrac{\lambda}{4} \end{cases}$$

Hence a quarter-wavelength line having an open-circuit load appears as a short circuit!

Can we construct other circuit elements with a short-circuited section of transmission line of different lengths? A short-circuited line has an input impedance of

$$\hat{Z}_{in} = Z_C \frac{1-e^{-j2\beta\mathscr{L}}}{1+e^{-j2\beta\mathscr{L}}} \begin{cases} \text{short-circuit load} \\ \hat{Z}_L = 0 \end{cases}$$

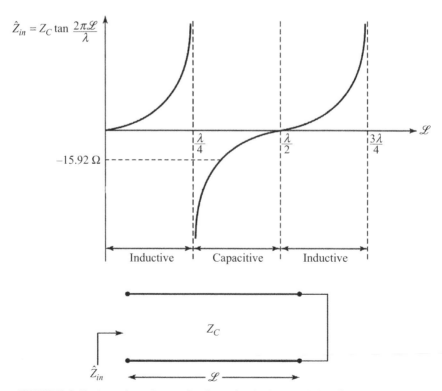

FIGURE 3.10. Input impedance of a short-circuited transmission line versus length.

But this can be written as

$$\hat{Z}_{in} = Z_C \frac{1 - e^{-j2\beta\mathscr{L}}}{1 + e^{-j2\beta\mathscr{L}}}$$

$$= Z_C \frac{e^{-j\beta\mathscr{L}}}{e^{-j\beta\mathscr{L}}} \left[\frac{e^{j\beta\mathscr{L}} - e^{-j\beta\mathscr{L}}}{e^{j\beta\mathscr{L}} + e^{-j\beta\mathscr{L}}} \right]$$

$$= jZ_C \tan\left(2\pi\frac{\mathscr{L}}{\lambda}\right) \begin{cases} \text{short-circuit load} \\ \hat{Z}_L = 0 \end{cases}$$

which can be plotted as shown in Fig. 3.10.

We can construct a 10-pF capacitor at 1 GHz using a short-circuited $Z_C = 50\text{-}\Omega$ line. The impedance of a 10-pF lumped-circuit capacitor at 1 GHz is

$$-j\frac{1}{\omega C} = -j15.92\,\Omega$$

Compare this to the equation for input impedance to a short-circuited transmission line:

$$jZ_C \tan\left(2\pi\frac{\mathscr{L}}{\lambda}\right) = -j15.92$$

Solve to give

$$\frac{\mathscr{L}}{\lambda} = -0.049$$

or

$$\frac{\mathscr{L}}{\lambda} = -0.049 + 0.5 = 0.451$$

If the line is air filled, $v = 3 \times 10^8$ m/s, the physical length (at 1 GHz) is

$$\mathscr{L} = 0.451 \times 0.3$$
$$= 13.53\,\text{cm}$$
$$= 5.33\,\text{in}$$

4

CROSSTALK IN THREE-CONDUCTOR LINES

Crosstalk is the inadvertent coupling of a signal on one transmission line onto an adjacent transmission line, thereby making possible interference in the modules that the adjacent line interconnects. In today's high-speed digital and high-frequency analog systems, this has become a critical design problem, preventing proper operation of those digital and analog systems. This is due to the electric fields $\vec{E}$ and magnetic fields $\vec{H}$ caused by the first line (the generator line) interacting with the second line (the receptor line), thereby inducing a voltage and a current in that line. The voltages induced at the ends of the receptor line are called the near-end and far-end crosstalk voltages, V_{NE} and V_{FE}, and can cause interference in the electronic devices that terminate the receptor line. In today's high-speed digital systems, crosstalk has become a very important factor in preventing them from working properly.

4.1 THE MULTICONDUCTOR TRANSMISSION-LINE EQUATIONS

Figure 4.1 shows the general configuration of a three-conductor transmission line that we will investigate. The *generator circuit* consists of the *loop* containing the source voltage waveform, $V_S(t)$, the source resistance, R_S,

Transmission Lines in Digital Systems for EMC Practitioners, First Edition.
Clayton R. Paul.
© 2012 John Wiley & Sons, Inc. Published 2012 by John Wiley & Sons, Inc.

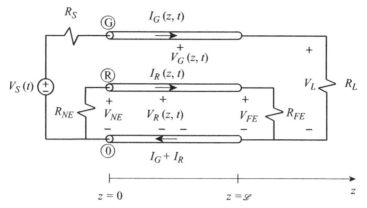

FIGURE 4.1. General three-conductor transmission line supporting crosstalk.

the generator conductor (denoted as G), the load resistance, R_L, and the reference conductor (denoted as 0). The *receptor circuit* consists of the *loop* containing the near-end resistance, R_{NE}, the receptor conductor (denoted as R), the far-end resistance, R_{FE}, and the reference conductor. The third conductor is referred to as the *reference conductor* and is denoted as the zeroth (0) conductor and serves as (1) the reference for the voltages of the generator circuit, $V_G(z, t)$, and the receptor circuit, $V_R(z, t)$, and (2) the *return* for the generator circuit current, $I_G(z, t)$, and the receptor circuit current, $I_R(z, t)$. Note that the current in the reference conductor is $I_0 = -(I_G + I_R)$, so that the net current along the system in the z direction is zero. All three conductors are assumed to be (1) parallel, (2) of uniform cross section along the line, and (3) of total length $\mathcal{L}$.

The current in the generator circuit, $I_G(z, t)$, creates a *magnetic field*, $\vec{H}_G(z, t)$, about it. This magnetic field threads the generator circuit giving a per-unit-length self-inductance of the generator circuit, l_G H/m. A portion of this magnetic field also threads the receptor circuit, thereby inducing, by Faraday's law, a current $I_R(z, t)$ around the receptor circuit. Hence a per-unit-length mutual inductance exists between the generator and receptor circuits, l_m H/m. Similarly, the current, $I_R(z, t)$, in the receptor circuit creates a *magnetic field*, $\vec{H}_R(z, t)$, about it that threads the receptor circuit, giving a per-unit-length self-inductance of the receptor circuit, l_R H/m. A portion of this magnetic field also threads the generator circuit which is again represented as a per-unit-length mutual inductance between the receptor and generator circuits, l_m H/m. (The mutual inductances between the two circuits are identical: $l_{GR} = l_{RG} \equiv l_m$.)

Similarly, the voltage of the generator circuit, $V_G(z, t)$, causes a per-unit-length charge, q_G, to be deposited on the generator circuit conductor which generates an *electric field*, $\vec{E}_G(z, t)$, between the conductors of the generator circuit and between the conductors of the generator and receptor circuits. This *induces* a per-unit-length charge, q_R, on the conductors of the receptor circuit, resulting in a voltage, $V_R(z, t)$, being induced between the conductors of the receptor circuit. These charges also create electric fields between the generator and receptor circuits. Hence we have per-unit-length self-capacitances of each circuit, c_G F/m and c_R F/m, along with a per-unit-length mutual capacitance, c_m F/m, between the two circuits. (The mutual capacitances between the two circuits are equal: $c_{GR} = c_{RG} \equiv c_m$.)

Figure 4.2 shows the cross sections of typical three-conductor transmission lines composed of *wires*: three wires, two wires above a ground plane, and two wires within a circular, cylindrical shield. The wires have radii r_w and have circular cylindrical insulations of thicknesses t about them having relative permittivities ε_r. In the first case, another wire serves as the reference conductor. In the second case an infinite ground plane serves

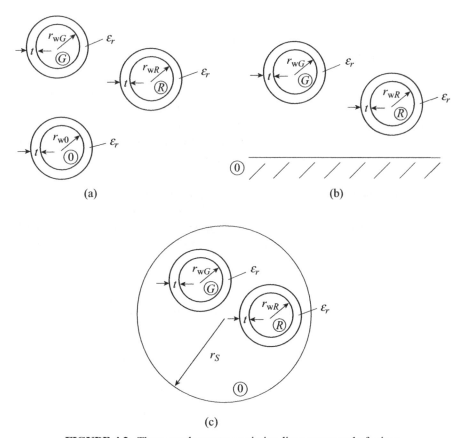

FIGURE 4.2. Three-conductor transmission lines composed of wires.

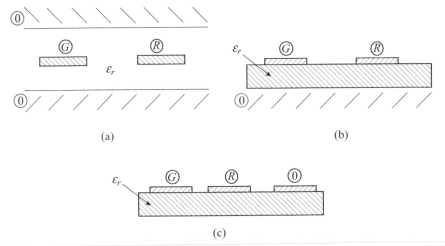

FIGURE 4.3. Three-conductor transmission lines composed of PCB lands.

as the reference conductor. In the third case an overall shield serves as the reference conductor.

Figure 4.3 shows the cross sections of three-conductor lines whose conductors have rectangular cross sections (lands) that are found on (and in) printed-circuit boards (PCBs). The first is a coupled stripline, the second is a coupled microstrip, and the third is an ordinary PCB not having inner planes. Note that the stripline is in a *homogeneous* medium, whereas the others are in an *inhomogeneous* medium since the electric field is partly in free space and partly in the dielectric substrate.

Again the electric and magnetic fields lie in the transverse plane perpendicular to the line z axis. Hence the mode of propagation along the line is the transverse electromagnetic (TEM) mode of propagation. The per-unit-length equivalent circuit is as shown in Fig. 4.4. Writing KVL around each loop gives

$$
\begin{aligned}
V_G(z+\Delta z, t) - V_G(z, t) &= -l_G\,\Delta z\,\frac{\partial I_G(z, t)}{\partial t} \\
&\quad -l_m\,\Delta z\,\frac{\partial I_R(z, t)}{\partial t} \\
V_R(z+\Delta z, t) - V_R(z, t) &= -l_m\,\Delta z\,\frac{\partial I_G(z, t)}{\partial t} \\
&\quad -l_R\,\Delta z\,\frac{\partial I_R(z, t)}{\partial t}
\end{aligned}
\tag{4.1a}
$$

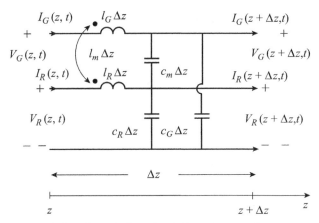

FIGURE 4.4. Per-unit-length equivalent circuit for a three-conductor transmission line.

Writing KCL at each upper node gives

$$I_G(z + \Delta z, t) - I_G(z, t) = -c_G \Delta z \frac{\partial V_G(z + \Delta z, t)}{\partial t}$$
$$- c_m \Delta z \frac{\partial (V_G(z + \Delta z, t) - V_R(z + \Delta z, t))}{\partial t}$$
$$I_R(z + \Delta z, t) - I_R(z, t) = -c_m \Delta z \frac{\partial (V_R(z + \Delta z, t) - V_G(z + \Delta z, t))}{\partial t}$$
$$- c_R \Delta z \frac{\partial V_R(z + \Delta z, t)}{\partial t} \tag{4.1b}$$

Dividing both sides by Δz and letting $\Delta z \rightarrow 0$ gives the *multiconductor transmission line (MTL) equations* as

$$\boxed{\begin{aligned}\frac{\partial V_G(z, t)}{\partial z} &= -l_G \frac{\partial I_G(z, t)}{\partial t} - l_m \frac{\partial I_R(z, t)}{\partial t} \\ \frac{\partial V_R(z, t)}{\partial z} &= -l_m \frac{\partial I_G(z, t)}{\partial t} - l_R \frac{\partial I_R(z, t)}{\partial t}\end{aligned}} \tag{4.2a}$$

and

$$\boxed{\begin{aligned}\frac{\partial I_G(z, t)}{\partial z} &= -(c_G + c_m) \frac{\partial V_G(z, t)}{\partial t} + c_m \frac{\partial V_R(z, t)}{\partial t} \\ \frac{\partial I_R(z, t)}{\partial z} &= c_m \frac{\partial V_G(z, t)}{\partial t} - (c_R + c_m) \frac{\partial V_R(z, t)}{\partial t}\end{aligned}} \tag{4.2b}$$

These are *coupled partial differential equations*. They can be written compactly in matrix form as

$$
\begin{aligned}
\frac{\partial}{\partial z}\mathbf{V}(z,t) &= -\mathbf{L}\frac{\partial}{\partial t}\mathbf{I}(z,t) \\
\frac{\partial}{\partial z}\mathbf{I}(z,t) &= -\mathbf{C}\frac{\partial}{\partial t}\mathbf{V}(z,t)
\end{aligned}
\tag{4.3}
$$

where the per-unit-length inductance and capacitance matrices are

$$
\mathbf{L} = \begin{bmatrix} l_G & l_m \\ l_m & l_R \end{bmatrix} \quad \frac{\text{H}}{\text{m}}
\tag{4.4a}
$$

$$
\mathbf{C} = \begin{bmatrix} c_G + c_m & -c_m \\ -c_m & c_R + c_m \end{bmatrix} \quad \frac{\text{F}}{\text{m}}
\tag{4.4b}
$$

and

$$
\mathbf{V}(z,t) = \begin{bmatrix} V_G(z,t) \\ V_R(z,t) \end{bmatrix}
$$

$$
\mathbf{I}(z,t) = \begin{bmatrix} I_G(z,t) \\ I_R(z,t) \end{bmatrix}
\tag{4.5}
$$

The uncoupled second-order MTL equations are

$$
\begin{aligned}
\frac{\partial^2}{\partial z^2}\mathbf{V}(z,t) &= \mathbf{LC}\frac{\partial^2}{\partial t^2}\mathbf{V}(z,t) \\
\frac{\partial^2}{\partial z^2}\mathbf{I}(z,t) &= \mathbf{CL}\frac{\partial^2}{\partial t^2}\mathbf{I}(z,t)
\end{aligned}
\tag{4.6}
$$

For MTLs the order of multiplication of $\mathbf{L}$ and $\mathbf{C}$ matters.

The line voltages and currents, $\mathbf{V}(z,t)$ and $\mathbf{I}(z,t)$, are functions of position along the line, z, and time, t, and are said to be in the *time domain*. The phasor, *frequency-domain* solution is obtained by exciting the line with a single-frequency sinusoidal source, $V_S(t) = V_S\cos(\omega t + \theta_S)$, giving similar forms of the line voltage and current as $V(z,t) = V\cos(\omega t + \theta_V)$ and $I(z,t) = I\cos(\omega t + \theta_I)$. These are symbolized as $\hat{V}_S = V_S\angle\theta_S$, $\hat{V}(z) = V(z)\angle\theta_V(z)$, and $\hat{I}(z) = I(z)\angle\theta_I(z)$, where these *phasor* quantities are complex numbers and are denoted as before with a caret and are functions only of position along the line, z. The phasor MTL equations are obtained by replacing all derivatives with respect to time t as $\partial/\partial t \Rightarrow j\omega$:

$$
\begin{aligned}
\frac{d}{dz}\hat{\mathbf{V}}(z) &= -j\omega\mathbf{L}\hat{\mathbf{I}}(z) \\
\frac{d}{dz}\hat{\mathbf{I}}(z) &= -j\omega\mathbf{C}\hat{\mathbf{V}}(z)
\end{aligned}
\tag{4.7}
$$

and the MTL equations become ordinary differential equations independent of time, t. The second-order uncoupled phasor differential equations become

$$
\begin{aligned}
\frac{d^2}{dz^2}\hat{\mathbf{V}}(z) &= -\omega^2\mathbf{L}\mathbf{C}\hat{\mathbf{V}}(z) \\
\frac{d^2}{dz^2}\hat{\mathbf{I}}(z) &= -\omega^2\mathbf{C}\mathbf{L}\hat{\mathbf{I}}(z)
\end{aligned}
\tag{4.8}
$$

The solution to these differential equations is a bit more tedious than the solution for two-conductor lines. We will avoid direct solution of these MTL equations but, instead, will develop a much simpler approximate solution in Chapter 5 that will give acceptable solutions for numerous practical cases. An *exact solution* of the time-domain MTL equations in (4.3) and (4.6) and the frequency-domain MTL equations in (4.7) and (4.8) is developed for a SPICE subcircuit model in Chapter 6.

4.2 THE MTL PER-UNIT-LENGTH PARAMETERS OF INDUCTANCE AND CAPACITANCE

The entries in the per-unit-length parameter matrices:

$$
\mathbf{L} = \begin{bmatrix} l_G & l_m \\ l_m & l_R \end{bmatrix} \quad \frac{\mathrm{H}}{\mathrm{m}}
\tag{4.4a}
$$

$$
\mathbf{C} = \begin{bmatrix} c_G + c_m & -c_m \\ -c_m & c_R + c_m \end{bmatrix} \quad \frac{\mathrm{F}}{\mathrm{m}}
\tag{4.4b}
$$

are determined in the same fashion as for two-conductor lines. However, for MTLs the calculations are more tedious and fall into two distinct categories.

> *Without the ability to compute the per-unit-length line parameters in (4.4), we*
> *CANNOT "solve" the MTL equations.*

So it is of critical importance that we determine numerical calculations for these per-unit-length parameters or our primary goal of "solving the MTL equations" will never be realized.

For wires, rather simple *approximate* results can easily be obtained, assuming that the wires are *widely separated*, and can be obtained as described in Section 4.2.1. This wide-separation approximation is similar to the assumption for two-wire lines where the ratio of wire separation to wire radii is on the order of 4 : 1 or greater, which is not very restrictive and results in about 5% error. In addition, it is assumed that the wires are in a *homogeneous medium* where we neglect the wire insulations. Like the case of two-conductor lines, $\mathbf{L}$ and $\mathbf{C}$ are related as

$$\mathbf{LC} = \mathbf{CL} = \underbrace{\mu\varepsilon}_{1/v^2} \mathbf{1}_2 \qquad \text{homogeneous medium} \qquad (4.9)$$

where $\mathbf{1}_2$ is the 2×2 identity matrix:

$$\mathbf{1}_2 = \begin{bmatrix} 1 & 0 \\ 0 & 1 \end{bmatrix}$$

and v is the velocity of propagation of the TEM waves along the line:

$$v = \frac{1}{\sqrt{\mu\varepsilon}}$$

Hence for lines in a *homogeneous medium*, each per-unit-length parameter matrix can be found from the other:

$$\mathbf{L} = \frac{1}{v^2}\mathbf{C}^{-1} \qquad \text{homogeneous medium}$$

and

$$\mathbf{C} = \frac{1}{v^2}\mathbf{L}^{-1} \qquad \text{homogeneous medium}$$

For this case, $\mathbf{L}$ and $\mathbf{C}$ are 2×2 matrices, and the inverse of a 2×2 matrix can easily be obtained by hand as

$$
\begin{aligned}
\mathbf{M}^{-1} &= \begin{bmatrix} a & b \\ c & d \end{bmatrix}^{-1} \\
&= \frac{1}{ad - bc} \begin{bmatrix} d & -b \\ -c & a \end{bmatrix}
\end{aligned}
$$

In other words, the inverse of a 2×2 matrix can be found by (1) swapping the main diagonal terms, (2) negating the off-diagonal terms, and (3) dividing each term by the determinant of the matrix, $ad - bc$. Usually, we first obtain the per-unit-length self-inductances l_G and l_R and the per-unit-length mutual inductance l_m by direct calculation. From $\mathbf{C} = (1/v^2)\mathbf{L}^{-1}$ for a *homogeneous medium*, we have the important identity

$$
\begin{aligned}
\mathbf{C} &= \frac{1}{v^2}\mathbf{L}^{-1} \\
&= \frac{1}{v^2} \frac{1}{l_G l_R - l_m^2} \begin{bmatrix} l_R & -l_m \\ -l_m & l_G \end{bmatrix} \\
&= \begin{bmatrix} c_G + c_m & -c_m \\ -c_m & c_R + c_m \end{bmatrix}
\end{aligned}
$$

By comparing terms we can then obtain the per-unit-length capacitances in terms of the per-unit-length inductances as

$$
c_m = \frac{1}{v^2} \frac{l_m}{l_G l_R - l_m^2} \tag{4.10a}
$$

$$
c_G + c_m = \frac{1}{v^2} \frac{l_R}{l_G l_R - l_m^2} \tag{4.10b}
$$

$$
c_R + c_m = \frac{1}{v^2} \frac{l_G}{l_G l_R - l_m^2} \tag{4.10c}
$$

The second category requires numerical methods and computer implementation and is described in Section 4.2.2. This applies to closely spaced wires with insulation such as ribbon cables and conductors of rectangular cross sections (lands) that are found on and in PCBs. Computer programs have been written by the author to compute **L** and **C** for these cases and can be downloaded from

ftp://ftp.wiley.com/public/sci_tech_med/multiconductor_transmission/

These are described in Section 4.2.2 and are RIBBON.FOR (RIBBON.EXE) for ribbon cables, PCB.FOR (PCB.EXE) for PCBs [shown in Fig. 4.3(c)], MSTRP.FOR (MSTRP.EXE) for coupled microstrip lines [shown in Fig. 4.3(b)], and STRPLINE.FOR (STRPLINE.EXE) for coupled striplines [shown in Fig. 4.3(a)]. The XXXX.FOR are the source codes written in FORTRAN, and XXXX.EXE are the corresponding compiled and executable codes. In addition, the codes WIDESEP.FOR (WIDESEP.EXE) for the wide-separation techniques for widely separated formulas in Section 4.2.1 are included, although the entries in **L** and **C** could be obtained straightforwardly from the formulas described in Section 4.2.1. Each of these computer programs requires only one input file describing the cross-sectional dimensions, XXXX. IN. The output of each program is delivered to the file PUL.DAT, which will contain the computed per-unit-length parameters in the upper triangles of **L** and **C**: l_G, l_R, l_m and c_G, c_R, c_m.

The per-unit-length inductance matrix, **L**, relates the per-unit-length magnetic fluxes that thread the generator and receptor circuit *loops* to the currents on the conductors as

$$\psi_G = l_G I_G + l_m I_R$$
$$\psi_R = l_m I_G + l_R I_R$$

(4.11a)

or

$$\boldsymbol{\psi} = \mathbf{L\,I}$$

(4.11b)

The per-unit-length capacitance matrix, **C**, relates the per-unit-length electric charges on the generator and receptor circuit conductors to the voltages of each circuit as

$$q_G = (c_G + c_m)V_G - c_m V_R$$
$$q_R = -c_m V_G + (c_R + c_m)V_R$$

(4.12a)

or

$$\mathbf{q} = \mathbf{C\,V}$$

(4.12b)

We can determine the individual per-unit-length parameters as

$$l_G = \left.\frac{\psi_G}{I_G}\right|_{I_R=0}$$

$$l_R = \left.\frac{\psi_R}{I_R}\right|_{I_G=0}$$

$$l_m = \left.\frac{\psi_G}{I_R}\right|_{I_G=0}$$

$$= \left.\frac{\psi_R}{I_G}\right|_{I_R=0}$$

(4.13a)

and

$$c_G + c_m = \left.\frac{q_G}{V_G}\right|_{V_R=0}$$

$$c_R + c_m = \left.\frac{q_R}{V_R}\right|_{V_G=0}$$

$$c_m = -\left.\frac{q_G}{V_R}\right|_{V_G=0}$$

$$= -\left.\frac{q_R}{V_G}\right|_{V_R=0}$$

(4.13b)

4.2.1   Wide-Separation Approximations for Wires

For these computations for wire-type lines we assume that (1) the wires are *widely separated* and (2) are immersed in a *homogeneous medium* that has parameters $\mu = \mu_0$ and $\varepsilon = \varepsilon_r \varepsilon_0$, where for free space (air), $\mu_0 = 4\pi \times 10^{-7}$ H/m and $\varepsilon_0 \cong (1/36\pi) \times 10^{-9}$ F/m. The restriction of the wires being widely separated is not as restrictive as it may seem. For wires that are very close together, the currents and charges on them will migrate toward the facing sides. This is called *proximity effect* and complicates the determination of the per-unit-length parameters. The wide-separation assumption is reasonably valid for close separations where the ratio of wire separation to wire radius is as small as 4 : 1. In other words, one wire would just fit between two adjacent wires. Hence, for practical applications this is not a very restrictive assumption.

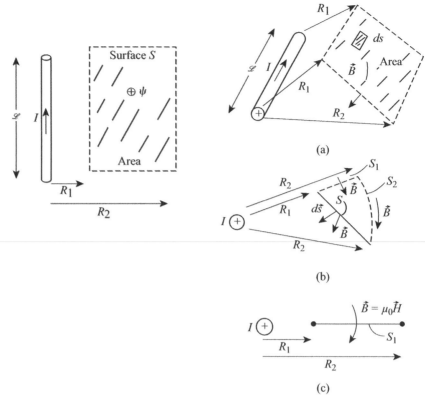

FIGURE 4.5. Fundamental subproblem for determining the per-unit-length inductances of wires.

For "widely separated" wires we may compute these per-unit-length parameters by assuming that the currents and the per-unit-length charges along the wires are distributed uniformly about the axes of the wires (i.e., the proximity effect is not well developed). Hence we can use the important results for the two fundamental subproblems developed earlier. The per-unit-length magnetic flux threading a surface that is parallel to the wire shown in Fig. 4.5 is

$$\psi = \frac{\mu_0 I}{2\pi} \ln \frac{R_2}{R_1} \quad \frac{\text{Wb}}{\text{m}} \qquad R_2 > R_1 \qquad (4.14)$$

The voltage between two points at distances R_1 and R_2 from the wire is obtained from Fig. 4.6 as

$$V = \frac{q}{2\pi\varepsilon} \ln \frac{R_2}{R_1} \quad \text{V} \qquad R_2 > R_1 \qquad (4.15)$$

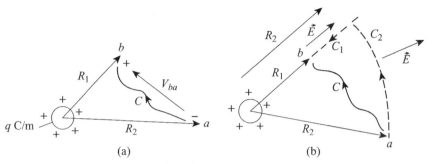

FIGURE 4.6. Fundamental subproblem for determining the per-unit-length capacitances of wires.

The first case is where the reference conductor is another wire. With reference to Fig. 4.7 and superimposing the fundamental subproblems above while assuming that the wires are "widely spaced" so that the currents and per-unit-length charge distributions are uniformly distributed around the wire peripheries, the entries in the per-unit-length inductance matrix are

$$
\begin{aligned}
l_G &= \left.\frac{\psi_G}{I_G}\right|_{I_R=0} \\
&= \frac{\mu_0}{2\pi}\ln\frac{d_G}{r_{w0}} + \frac{\mu_0}{2\pi}\ln\frac{d_G}{r_{wG}} \\
&= \frac{\mu_0}{2\pi}\ln\frac{d_G^2}{r_{w0}r_{wG}}
\end{aligned}
\tag{4.16a}
$$

$$
\begin{aligned}
l_R &= \left.\frac{\psi_R}{I_R}\right|_{I_G=0} \\
&= \frac{\mu_0}{2\pi}\ln\frac{d_R}{r_{w0}} + \frac{\mu_0}{2\pi}\ln\frac{d_R}{r_{wR}} \\
&= \frac{\mu_0}{2\pi}\ln\frac{d_R^2}{r_{w0}r_{wR}}
\end{aligned}
\tag{4.16b}
$$

$$
\begin{aligned}
l_m &= \left.\frac{\psi_G}{I_R}\right|_{I_G=0} = \left.\frac{\psi_R}{I_G}\right|_{I_R=0} \\
&= \frac{\mu_0}{2\pi}\ln\frac{d_R}{d_{GR}} + \frac{\mu_0}{2\pi}\ln\frac{d_G}{r_{w0}} \\
&= \frac{\mu_0}{2\pi}\ln\frac{d_G d_R}{d_{GR}r_{w0}}
\end{aligned}
\tag{4.16c}
$$

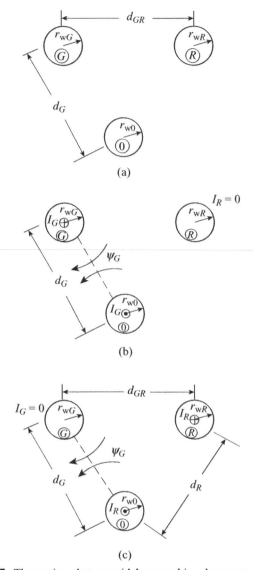

FIGURE 4.7. Three wires that are widely spaced in a homogeneous medium.

In the case of a homogeneous medium, we can determine the per-unit-length capacitances from $\mathbf{C} = (1/v^2)\mathbf{L}^{-1}$, where the propagation velocity in this *homogeneous medium* characterized by μ_0 and $\varepsilon = \varepsilon_r\varepsilon_0$ is $v = 1/\sqrt{\mu_0\varepsilon_r\varepsilon_0}$.

The next case is for two wires above an infinite perfectly conducting ground plane. The ground plane is the reference conductor. We solve this problem by replacing the infinite perfectly conducting ground plane with its image. With reference to Fig. 4.8 and superimposing the fundamental

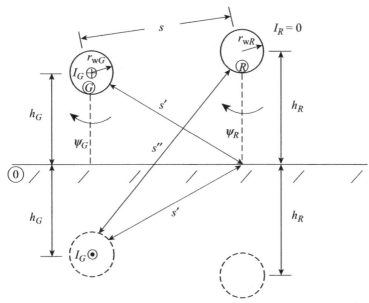

FIGURE 4.8. Two wires widely spaced in a homogeneous medium above an infinite ground plane.

subproblems above while assuming that the wires are "widely spaced" so that the currents and per-unit-length charge distributions are uniformly distributed around the wire peripheries, the entries in the per-unit-length inductance matrix are

$$
\begin{aligned}
l_G &= \left.\frac{\psi_G}{I_G}\right|_{I_R=0} \\
&= \frac{\mu_0}{2\pi}\ln\frac{h_G}{r_{wG}} + \frac{\mu_0}{2\pi}\ln\frac{2h_G}{h_G} \\
&= \frac{\mu_0}{2\pi}\ln\frac{2h_G}{r_{wG}}
\end{aligned}
\tag{4.17a}
$$

$$
\begin{aligned}
l_R &= \left.\frac{\psi_R}{I_R}\right|_{I_G=0} \\
&= \frac{\mu_0}{2\pi}\ln\frac{h_R}{r_{wR}} + \frac{\mu_0}{2\pi}\ln\frac{2h_R}{h_R} \\
&= \frac{\mu_0}{2\pi}\ln\frac{2h_R}{r_{wR}}
\end{aligned}
\tag{4.17b}
$$

and

$$
\begin{aligned}
l_m &= \frac{\psi_R}{I_G}\bigg|_{I_R=0} = \frac{\psi_G}{I_R}\bigg|_{I_G=0} \\[2mm]
&= \frac{\mu_0}{2\pi}\ln\frac{s'}{s} + \frac{\mu_0}{2\pi}\ln\frac{s''}{s'} \\[2mm]
&= \frac{\mu_0}{2\pi}\ln\frac{\sqrt{s^2 + 4h_G h_R}}{s^2} \\[2mm]
&= \frac{\mu_0}{4\pi}\ln\left(1 + \frac{4h_G h_R}{s^2}\right)
\end{aligned}
\tag{4.17c}
$$

In the case of a homogeneous medium, we can determine the per-unit-length capacitances from $\mathbf{C} = (1/v^2)\mathbf{L}^{-1}$ where the propagation velocity in this *homogeneous medium* characterized by μ_0 and $\varepsilon = \varepsilon_r \varepsilon_0$ is $v = 1/\sqrt{\mu_0 \varepsilon_r \varepsilon_0}$.

The final case is for two wires located within an overall circular cylindrical shield. The shield is the reference conductor. The shield can be replaced with images located at distances r_s^2/d_G and r_s^2/d_R. With reference to Fig. 4.9 and superimposing the fundamental subproblems above while assuming that the wires are "widely spaced" so that the currents and per-unit-length charge distributions are uniformly distributed around the wire peripheries, the entries in the per-unit-length inductance matrix are

$$
\begin{aligned}
l_G &= \frac{\psi_G}{I_G}\bigg|_{I_R=0} \\[2mm]
&= \frac{\mu_0}{2\pi}\ln\frac{r_s - d_G}{r_{wG}} + \frac{\mu_0}{2\pi}\ln\frac{(r_s^2/d_G) - d_G}{(r_s^2/d_G) - r_s} \\[2mm]
&= \frac{\mu_0}{2\pi}\ln\frac{r_s^2 - d_G^2}{r_s r_{wG}}
\end{aligned}
\tag{4.18a}
$$

$$
\begin{aligned}
l_R &= \frac{\psi_R}{I_R}\bigg|_{I_G=0} \\[2mm]
&= \frac{\mu_0}{2\pi}\ln\frac{r_s - d_R}{r_{wR}} + \frac{\mu_0}{2\pi}\ln\left(\frac{r_s^2/d_R - d_R}{r_s^2/d_R - r_s}\right) \\[2mm]
&= \frac{\mu_0}{2\pi}\ln\frac{r_s^2 - d_R^2}{r_s r_{wR}}
\end{aligned}
\tag{4.18b}
$$

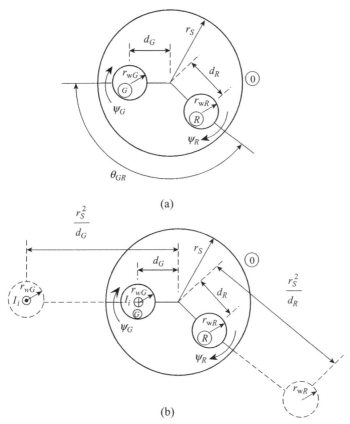

FIGURE 4.9. Two wires widely spaced in a homogeneous medium within an overall cylindrical shield.

$$
\begin{aligned}
l_m &= \left.\frac{\psi_R}{I_G}\right|_{I_R=0} = \left.\frac{\psi_G}{I_R}\right|_{I_G=0} \\
&= \frac{\mu_0}{2\pi} \ln\left[\frac{d_R}{r_s}\sqrt{\frac{(d_G d_R)^2 + r_s^4 - 2d_G d_R r_s^2 \cos\theta_{GR}}{(d_G d_R)^2 + d_R^4 - 2d_G d_R^3 \cos\theta_{GR}}}\right]
\end{aligned}
$$

(4.18c)

In the case of a homogeneous medium, we can determine the per-unit-length capacitances from $\mathbf{C} = (1/v^2)\mathbf{L}^{-1}$, where the propagation velocity in this *homogeneous medium* characterized by μ_0 and $\varepsilon = \varepsilon_r \varepsilon_0$ is $v = 1/\sqrt{\mu_0 \varepsilon_r \varepsilon_0}$.

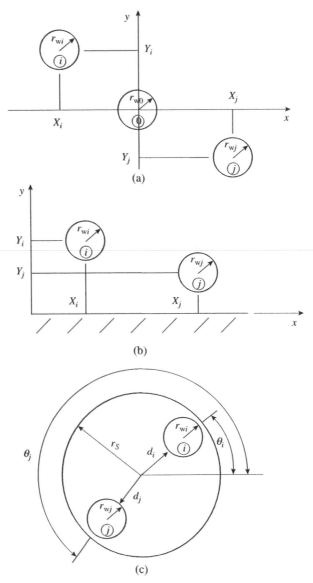

FIGURE 4.10. Descriptions of the locations of the wires for the WIDESEP. FOR program used in the input file WIDESEP.IN.

The WIDESEP Computer Program (WIDESEP.FOR, WIDESEP.EXE)
WIDESEP.FOR is a FORTRAN program developed to compute the per-unit-length **L** and **C** matrices for widely separated wires. The program WIDESEP.EXE is the compiled and executable file. WIDESEP.EXE reads the input data stored in the ASCII file WIDESEP.IN and produces as output the file

PUL.DAT, which contains the upper diagonals of the per-unit-length matrices **L** and **C**. Although our concentration will be on three-conductor lines, the program handles n conductors, where n is any number up to the array dimensions of WIDESEP.FOR.

The locations of the wires for (1) three wires and (2) two wires above a ground plane are specified in a rectangular x–y coordinate system as shown in Fig. 4.10 (a) and (b). For three wires, the reference wire is located at the origin of the coordinate system. For the case of two wires above a ground plane, the ground plane is the x axis of the coordinate system. In the case of two wires within an overall circular, cylindrical shield, the locations of the wires are specified in terms of the distance of the wire from the center of the shield and its angular location with reference to Fig. 4.10(c).

Following are three examples of the use of the WIDESEP.FOR (WIDESEP. EXE) program to compute the entries in the per-unit-length inductance and capacitance matrices **L** and **C** that are the upper diagonals output to the file PUL.DAT. Note that any such data overwrite any previous data written to the file PUL.DAT. There is only one input file to WIDESEP.FOR (WIDESEP. EXE) and it is is prepared using a standard ASCII editor and is called WIDESEP.IN.

As an example, the WIDESEP.IN data for three wires are

```
2           =NUMBER OF WIRES (EXCLUDING REFERENCE CONDUCTOR)
1           =REF CONDUCTOR (1= WIRE, 2= GND PLANE, 3= OVERALL SHIELD)
1.0         =RELATIVE PERMITTIVITY ER OF HOMOGENEOUS MEDIUM
1.0         =RELATIVE PERMEABILITY MUR OF HOMOGENEOUS MEDIUM
7.5         =REF WIRE RADIUS OR SHIELD INTERIOR RADIUS (mils)
1           =WIRE #1
7.5         =RADIUS OF WIRE #1 (mils)
-1.27E-3 =1= X1(meters), 2= H1(meters), 3= D1(mils)
0.          =1= Y1(meters), 2= Y1(meters), 3= THETA1(degrees)
2           =WIRE #2
7.5         =RADIUS OF WIRE #2 (mils)
1.27E-3  =1= X2(meters), 2= H2(meters), 3= D2(mils)
0.          =1= Y2(meters), 2= Y2(meters), 3= THETA2(degrees)
```

The output data are in PUL.DAT:

```
1   1   7.58848E-07    =L( 1, 1)
1   2   2.40795E-07    =L( 1, 2)
2   2   7.58848E-07    =L( 2, 2)
1   1   1.63040E-11    =C( 1, 1)
```

```
1   2  -5.17352E-12    =C( 1, 2)
2   2   1.63040E-11    =C( 2, 2)
NUMBER OF WIRES= 2
RELATIVE PERMITTIVITY OF HOMOGENEOUS MEDIUM= 1.0
RELATIVE PERMEABILITY OF HOMOGENEOUS MEDIUM= 1.0
REFERENCE CONDUCTOR IS A WIRE WITH RADIUS (mils)= 7.500E+00
```

WIRE #	WIRE RADIUS (mils)	X COORDINATE (meters)	Y COORDINATE (meters)
1	7.500E+00	-1.270E-03	0.000E+00
2	7.500E+00	1.270E-03	0.000E+00

The WIDESEP.IN data for two wires above a ground plane are

```
2       = NUMBER OF WIRES (EXCLUDING REFERENCE CONDUCTOR)
2       = REF CONDUCTOR (1=WIRE, 2=GND PLANE, 3=OVERALL SHIELD)
1.0     = RELATIVE PERMITTIVITY ER OF HOMOGENEOUS MEDIUM
1.0     = RELATIVE PERMEABILITY MUR OF HOMOGENEOUS MEDIUM
16.     = REF WIRE RADIUS OR SHIELD INTERIOR RADIUS (mils)
1       = WIRE #1
16.     = RADIUS OF WIRE #1 (mils)
2.E-2 = 1=X1(meters), 2=H1(meters), 3=D1(mils)
0.    = 1=Y1(meters), 2=Y1(meters), 3=THETA1(degrees)
2       = WIRE #2
16.     = RADIUS OF WIRE #2 (mils)
2.E-2 = 1=X2(meters), 2=H2(meters), 3=D2(mils)
2.E-2 = 1=Y2(meters), 2=Y2(meters), 3=THETA2(degrees)
```

The output data are in PUL.DAT:

```
1   1   9.17859E-07    =L( 1, 1)
1   2   1.60944E-07    =L( 1, 2)
2   2   9.17859E-07    =L( 2, 2)
1   1   1.25068E-11    =C( 1, 1)
1   2  -2.19302E-12    =C( 1, 2)
2   2   1.25068E-11    =C( 2, 2)

NUMBER OF WIRES= 2
RELATIVE PERMITTIVITY OF HOMOGENEOUS MEDIUM= 1.0
RELATIVE PERMEABILITY OF HOMOGENEOUS MEDIUM= 1.0
REFERENCE CONDUCTOR IS A GROUND PLANE
```

WIRE #	WIRE RADIUS (mils)	Y COORDINATE (meters)	HEIGHT ABOVE GND (meters)
1	1.600E+01	0.000E+00	2.000E-02
2	1.600E+01	2.000E-02	2.000E-02

The WIDESEP.IN data for two wires within an overall shield are

```
2      =NUMBER OF WIRES (EXCLUDING REFERENCE CONDUCTOR)
3      =REF CONDUCTOR (1=WIRE,2=GND PLANE,3=OVERALL SHIELD)
1.0    =RELATIVE PERMITTIVITY ER OF HOMOGENEOUS MEDIUM
1.0    =RELATIVE PERMEABILITY MUR OF HOMOGENEOUS MEDIUM
30.    =REF WIRE RADIUS OR SHIELD INTERIOR RADIUS (mils)
1      =WIRE #1
7.5    =RADIUS OF WIRE #1 (mils)
15.    =1=X1(meters), 2=H1(meters), 3=D1(mils)
0.     =1=Y1(meters), 2=Y1(meters), 3=THETA1(degrees)
2      =WIRE #2
7.5    =RADIUS OF WIRE #2 (mils)
15.    =1=X2(meters), 2=H2(meters), 3=D2(mils)
180.   =1=Y2(meters), 2=Y2(meters), 3=THETA2(degrees)
```

The output data are in PUL.DAT:

```
1   1    2.19722E-07     =L( 1, 1)
1   2    4.46287E-08     =L( 1, 2)
2   2    2.19722E-07     =L( 2, 2)
1   1    5.28179E-11     =C( 1, 1)
1   2   -1.07281E-11     =C( 1, 2)
2   2    5.28179E-11     =C( 2, 2)
```

```
NUMBER OF WIRES= 2
RELATIVE PERMITTIVITY OF HOMOGENEOUS MEDIUM= 1.0
RELATIVE PERMEABILITY OF HOMOGENEOUS MEDIUM= 1.0
REFERENCE CONDUCTOR IS A SHIELD WITH RADIUS (mils) = 3.000E+01
```

WIRE #	WIRE RADIUS (mils)	POSITION RADIUS (mils)	POSITION ANGLE (degrees)
1	7.500E+00	1.500E+01	0.000E+00
2	7.500E+00	1.500E+01	1.800E+02

4.2.2   Numerical Methods

Conductors Having Circular Cylindrical Cross Sections (Ribbon Cables)
Ribbon cables are flat arrays of identical wires with insulations that have
identical spacings (typically, 50 mils), as illustrated in Fig. 4.11. Typical

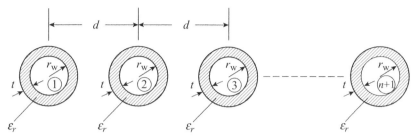

FIGURE 4.11. Common ribbon cable.

dimensions and properties of common ribbon cables are

$$d = 50 \, \text{mils}$$
$$r_w = 7.5 \, \text{mils (No. 20 gauge, } 7 \times 36)$$
$$t = 10 \, \text{mils}$$
$$\varepsilon_r = 3.5 \, (\text{PVC})$$

There are two important properties that prevent using the previous wide-separation formulas:

1. The metallic wires are surrounded by dielectric insulations. Hence this is an inhomogeneous medium.
2. The ratio of separation to overall radius of the wires (wire radius + dielectric thickness) is 2.86, which means that proximity effect will be pronounced and we cannot make the assumption that the per-unit-length charges will be distributed uniformly around the wire and dielectric peripheries.

To handle these problems, we note that the dielectric insulations will have *bound charge distributions* around their inner *and* outer surfaces, $\rho_b \, \text{C/m}^2$, while the surfaces of the metallic wires will have *free charge distributions* around them, $\rho_f \, \text{C/m}^2$. (One dimension is along the wire and the other is around the wire periphery.)

As an example, consider the parallel-plate capacitor illustrated in Fig. 4.12. When we attach the voltage source, *free charge* will flow from the battery to the capacitor plates. When we insert a block of dielectric between the plates, the microscopic *dipoles* in the dielectric will align with the $\vec{E}$ field, inducing *bound charges* on the faces of the dielectric. This will cause more *free charge* to flow from the battery to the plates. Hence the capacitance (which is the ratio of *free charge* to the voltage between the two plates) will increase.

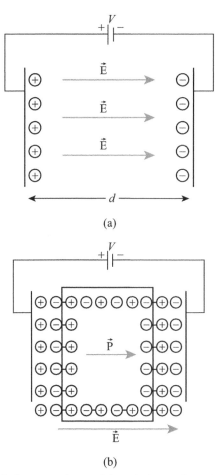

FIGURE 4.12. Parallel-plate capacitor and bound charge formed on the surfaces of the dielectric.

For the ribbon cable, we expand these distributions as Fourier series in terms of the angle around the surfaces. Around the outer surface of the dielectric insulation we represent the bound charge on its surface with a Fourier series as

$$\hat{\rho}_{ib} = \hat{\alpha}_{i0} + \sum_{k=1}^{\hat{N}_i} \hat{\alpha}_{ik} \cos(k\theta_i) \qquad \frac{C}{m^2} \qquad (4.19)$$

where the subscript i is for the ith wire. Around the outer surface of the metallic wire (the inner surface of the dielectric insulation) we have both free and bound charge. The *total* bound charge on the inner and outer surfaces of the dielectric insulation will be equal but opposite in sign. Because of the different radii of the surfaces, their distributions will be different. Similarly, we represent the bound and free charge on the conductor–dielectric surface with a Fourier series as

$$\rho_{if} - \rho_{ib} = \alpha_{i0} + \sum_{k=1}^{N_i} \alpha_{ik} \cos(k\theta_i) \qquad \frac{C}{m^2} \qquad (4.20)$$

At the air–dielectric surface, the total per-unit-length bound charge is

$$q_{ib} = \int_{\theta_i=0}^{2\pi} \hat{\rho}_{ib}(r_w + t)d\theta_i$$

$$= 2\pi(r_w + t)\hat{\alpha}_{i0} \qquad \frac{C}{m} \qquad (4.21)$$

At the conductor–dielectric surface, the total per-unit-length free plus bound charge is

$$q_{if} - q_{ib} = \int_{\theta_i=0}^{2\pi} (\rho_{if} - \rho_{ib})r_w d\theta_i$$

$$= 2\pi r_w \alpha_{i0} \qquad \frac{C}{m} \qquad (4.22)$$

Hence the total per-unit-length free charge around the metallic wire is

$$\boxed{q_{if} = 2\pi(r_w + t)\hat{\alpha}_{i0} + 2\pi r_w \alpha_{i0} \qquad \frac{C}{m}} \qquad (4.23)$$

Each wire has a total of $N_i + 1 + \hat{N}_i + 1$ unknown expansion coefficients associated with it. To determine these, we select $N_i + 1$ match points around the wire periphery, where we impose the condition that the total voltage there due to all the charge distribtutions will equal the voltage of that wire. We select $\hat{N}_i + 1$ match points around the air–dielectric periphery where we impose the condition that the components of the $\vec{D}$ fields due to all the charge distributions on either side of the boundary that are normal to the boundary will be equal. This gives a set of simultaneous equations to be solved for the α_{i0} and $\hat{\alpha}_{i0}$ average value coefficients for each wire, giving the per-unit-length capacitance matrix, C, for a chosen reference conductor. The per-unit-length inductance matrix can be found from the capacitance matrix with the dielectric insulations removed, C_0, as $L = (1/v_0^2)C_0^{-1}$.

The RIBBON Computer Program (RIBBON.FOR, RIBBON.EXE) RIBBON.FOR is a FORTRAN program developed to compute the per-unit-length **L** and **C** matrices for ribbon cables. The program RIBBON. EXE is the compiled and executable file. RIBBON.EXE reads the input data stored in the file RIBBON.IN and produces as output the file PUL. DAT, which contains the upper diagonals of the per-unit-length matrices **L** and **C**. (Note that any data in a previous PUL.DAT are overwritten by the new data.) Although our concentration will be on three-conductor lines, the program handles $n + 1$ wires, where $n + 1$ is any number up to the array dimensions of RIBBON.FOR. The number of the reference wire is chosen in RIBBON.IN. Note that the wires are numbered *left to right* as $1, 2, \ldots, n + 1$. When the reference wire is chosen (0), the wires are again numbered sequentially *from left to right*, omitting the chosen reference wire, as $1, 2, \ldots, n$.

For example, the RIBBON.IN data for three wires are

```
3            =TOTAL NUMBER OF WIRES
20           =NUMBER OF FOURIER COEFFICIENTS
2            =NUMBER OF REFERENCE WIRE
7.5          =WIRE RADIUS (mils)
10.0         =INSULATION THICKNESS (mils)
3.5          =RELATIVE DIELECTRIC CONSTANT OF INSULATION
50.0         =ADJACENT WIRE SEPARATION (mils)
```

The output data are in PUL.DAT:

```
1      1      7.48501E-07      =L ( 1, 1)
1      2      2.40801E-07      =L ( 1, 2)
2      2      7.48501E-07      =L ( 2, 2)
1      1      2.49819E-11      =C ( 1, 1)
1      2     -6.26613E-12      =C ( 1, 2)
2      2      2.49819E-11      =C ( 2, 2)
1      1      1.65812E-11      =C0 ( 1, 1)
1      2     -5.33434E-12      =C0 ( 1, 2)
2      2      1.65812E-11      =C0 ( 2, 2)
```

```
NUMBER OF WIRES= 3
NUMBER OF FOURIER COEFFICIENTS= 20
REFERENCE WIRE= 2
WIRE RADIUS (mils)= 7.500E+00
DIELECTRIC INSULATION THICKNESS (mils)= 1.000E+01
DIELECTRIC CONSTANT OF INSULATION= 3.500E+00
CENTER-TO-CENTER SEPARATION (mils)= 5.000E+01
```

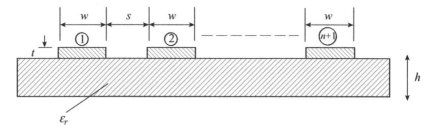

FIGURE 4.13. Land numbering for the PCB.FOR program.

Conductors Having Rectangular Cross Sections (PCB Lands) The conductors (lands) of PCBs have rectangular cross sections. The thickness of typical lands (etched from 1-ounce copper) is 1.4 mils, and the typical land widths range from 5 to 25 mils. To simplify the calculation, we assume that the land thicknesses are zero ($t = 0$). Typical thicknesses of the boards are 47 and 64 mils. The supporting board is made of glass epoxy (FR-4) and has a relative permittivity of around $\varepsilon_r = 4.7$. Again note that the lands are numbered *left to right* as $1, 2, \ldots, n + 1$, as illustrated in Fig. 4.13. When the reference land is chosen (0), the lands are again numbered *from left to right*, omitting the chosen reference land as $1, 2, \ldots, n$.

Numerical methods must again be used to compute the per-unit-length inductances and capacitances of PCB lands because the charge and current tends to *peak at the edges of the land cross sections*, as illustrated in Fig. 4.14. To determine the per-unit-length charge on each land, we model the charge distribution with piecewise-constant segments whose heights, α_{ik} C/m^2, are

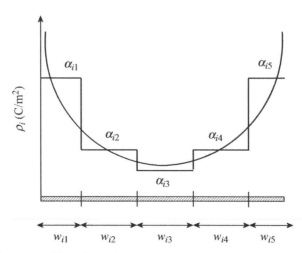

FIGURE 4.14. Peaking of the charge distributions across the lands at the edges.

unknown. Then the per-unit-length charge on the ith land is

$$q_i = \sum_{k=1}^{N_i} \alpha_{ik} w_{ik} \qquad \frac{C}{m} \qquad (4.24)$$

To determine the unknown distribution coefficients, α_{ik}, we again choose N_i match points at the midpoints of each of the subsections of each land at which we enforce the requirement that the voltage at the midpoint of that subsection that is due to all the charges on all the land subsections is equal to the voltage of the land. This again gives $(n+1)N_i$ equations in the $(n+1)N_i$ unknowns. We incorporate the inhomogeneous medium by incorporating the board into this, using images. Solving these for the α_{ik}, we can determine the per-unit-length capacitance matrix with $(\mathbf{C})$ and without $(\mathbf{C_0})$ the board present. Then the per-unit-length inductance matrix can be found from $\mathbf{L} = (1/v_0^2)\mathbf{C_0}^{-1}$.

The PCB Computer Program (PCB.FOR, PCB.EXE)   PCB.FOR is a FORTRAN program developed to compute the per-unit-length $\mathbf{L}$ and $\mathbf{C}$ matrices for the PCB shown in Fig. 4.13. The program PCB.EXE is the compiled and executable file. PCB.EXE reads the input data stored in the file PCB.IN and produces as output the file PUL.DAT, which contains the upper diagonals of the per-unit-length matrices $\mathbf{L}$ and $\mathbf{C}$. (Note that any data in a previous PUL.DAT are overwritten by the new data.) Although our concentration will be on three-conductor lines, the program handles $n+1$ conductors, where $n+1$ is any number up to the array dimensions of PCB.FOR.

As an example, the PCB.IN data for a three-land PCB are:

```
3            =TOTAL NUMBER OF LANDS
30           =NUMBER OF CONDUCTOR SUBSECTIONS
3            =NUMBER OF REFERENCE LAND
15.0         =LAND WIDTH (mils)
45.0         =EDGE-TO-EDGE LAND SEPARATION (mils)
47.          =BOARD THICKNESS (mils)
4.7          =BOARD RELATIVE DIELECTRIC CONSTANT
```

The output data are in PUL.DAT:

```
1    1    1.38314E-06    =L( 1, 1)
1    2    6.91570E-07    =L( 1, 2)
2    2    1.10706E-06    =L( 2, 2)
1    1    2.96949E-11    =C( 1, 1)
```

```
1     2     -2.02619E-11        =C ( 1, 2)
2     2      4.05238E-11        =C ( 2, 2)
1     1      1.16983E-11        =CO ( 1, 1)
1     2     -7.30777E-12        =CO ( 1, 2)
2     2      1.46155E-11        =CO ( 2, 2)


NUMBER OF LANDS= 3
NUMBER OF DIVISIONS PER LAND= 30
REFERENCE LAND= 3
LAND WIDTH (mils)= 1.500E+01
EDGE-TO-EDGE SEPARATION (mils)= 4.500E+01
BOARD THICKNESS (mils)= 4.700E+01
RELATIVE DIELECTRIC CONSTANT= 4.700E+00
```

The MSTRP Computer Program (MSTRP.FOR, MSTRP.EXE) MSTRP.
FOR is a FORTRAN program developed to compute the per-unit-length **L** and
C matrices for the coupled microstrip line shown in Fig. 4.3(b), where there is
a ground plane (the reference conductor) beneath the board. The program
MSTRP.EXE is the compiled and executable file. MSTRP.EXE reads the input
data stored in the file MSTRP.IN and produces as output the file PUL.DAT,
which contains the upper diagonals of the per-unit-length matrices **L** and **C**.
(Note that any data in a previous PUL.DAT are overwritten by the new data.)
Although our concentration will be on three-conductor lines (two lands and
the ground plane), the program handles *n* conductors, where *n* is any number
up to the array dimensions of MSTRP.FOR. The ground plane and the
substrate are included with the method of images.

**As an example, the MSTRP.IN data for a two-land line above a ground
plane coupled microstrip are:**

```
2              =TOTAL NUMBER OF LANDS (EXCLUSIVE OF GND PLANE)
30             =NUMBER OF CONDUCTOR SUBSECTIONS
100.           =LAND WIDTH (mils)
100.           =EDGE-TO-EDGE LAND SEPARATION (mils)
62.            =BOARD THICKNESS (mils)
4.7            =BOARD RELATIVE DIELECTRIC CONSTANT
```

The output data are in PUL.DAT:

```
1     1     3.35327E-07        =L ( 1, 1)
1     2     3.71527E-08        =L ( 1, 2)
2     2     3.35327E-07        =L ( 2, 2)
1     1     1.15511E-10        =C ( 1, 1)
```

```
1    2   -4.92518E-12         =C ( 1, 2)
2    2    1.15511E-10         =C ( 2, 2)
1    1    3.35935E-11         =C0 ( 1, 1)
1    2   -3.72200E-12         =C0 ( 1, 2)
2    2    3.35934E-11         =C0 ( 2, 2)


NUMBER OF LANDS= 2
NUMBER OF DIVISIONS PER LAND= 30
LAND WIDTH (mils)= 1.000E+02
EDGE-TO-EDGE SEPARATION (mils)= 1.000E+02
BOARD THICKNESS (mils)= 6.200E+01
RELATIVE DIELECTRIC CONSTANT= 4.700E+00
```

The STRPLINE Computer Program (STRPLINE.FOR, STRPLINE.EXE)

STRPLINE.FOR is a FORTRAN program developed to compute the per-unit-length L and C matrices for the coupled stripline shown in Fig. 4.3(a), where there is a ground plane (the reference conductor) on both sides of the board and the lands are sandwiched between them. Note that unlike the ribbon cable, the PCB, and the coupled microstrip line, this is a line in a *homogeneous medium*. The program STRPLINE.EXE is the compiled and executable file. STRPLINE.EXE reads the input data stored in the file STRPLINE.IN and produces as output the file PUL.DAT, which contains the upper diagonals of the per-unit-length matrices L and C. (Note that any data in a previous PUL. DAT are overwritten by the new data.) Although our concentration will be on three-conductor lines (two lands and the ground planes), the program handles n conductors, where n is any number up to the array dimensions of STRPLINE.FOR. The ground planes are included with the method of images.

As an example, the STRPLINE.IN data for a two-land line between ground planes coupled stripline are:

```
2              =TOTAL NUMBER OF LANDS (EXCLUSIVE OF GND PLANES)
30             =NUMBER OF CONDUCTOR SUBSECTIONS
5.             =LAND WIDTH (mils)
5.             =EDGE-TO-EDGE LAND SEPARATION (mils)
20.            =SEPARATION BETWEEN GROUND PLANES (mils)
4.7            =BOARD RELATIVE DIELECTRIC CONSTANT
```

The output data are in PUL.DAT:

```
1    1    4.65828E-07         =L ( 1, 1)
1    2    9.20798E-08         =L ( 1, 2)
```

2	2	4.65828E-07	=L (2, 2)
1	1	1.16826E-10	=C (1, 1)
1	2	-2.30930E-11	=C (1, 2)
2	2	1.16826E-10	=C (2, 2)
1	1	2.48567E-11	=C0 (1, 1)
1	2	-4.91340E-12	=C0 (1, 2)
2	2	2.48567E-11	=C0 (2, 2)

NUMBER OF LANDS= 2
NUMBER OF DIVISIONS PER LAND= 30
LAND WIDTH (mils) = 5.000E+00
EDGE-TO-EDGE SEPARATION (mils) = 5.000E+00
SEPARATION BETWEEN GROUND PLANES (mils) = 2.000E+01
RELATIVE DIELECTRIC CONSTANT= 4.700E+00

5

THE APPROXIMATE INDUCTIVE–CAPACITIVE CROSSTALK MODEL

In this chapter we develop a simple but approximate model of coupled three-conductor lines for predicting the crosstalk between them. Not only does this simple model give reasonably accurate approximate results for the crosstalk, but it also gives a considerable amount of *insight* into how the crosstalk is created. It also shows how such remedies as shielded wires and twisted pairs of wire can be used to affect that crosstalk.

The multiconductor transmission line (MTL) equations are, again, coupled partial differential equations. They can be written compactly in matrix form as

$$
\frac{\partial}{\partial z} \mathbf{V}(z, t) = -\mathbf{L} \frac{\partial}{\partial t} \mathbf{I}(z, t)
$$
$$
\frac{\partial}{\partial z} \mathbf{I}(z, t) = -\mathbf{C} \frac{\partial}{\partial t} \mathbf{V}(z, t)
$$

(5.1)

where the per-unit-length inductance and capacitance matrices are

$$
\mathbf{L} = \begin{bmatrix} l_G & l_m \\ l_m & l_R \end{bmatrix} \quad \frac{\mathrm{H}}{\mathrm{m}}
$$

(5.2a)

Transmission Lines in Digital Systems for EMC Practitioners, First Edition.
Clayton R. Paul.
© 2012 John Wiley & Sons, Inc. Published 2012 by John Wiley & Sons, Inc.

$$\mathbf{C} = \begin{bmatrix} c_G + c_m & -c_m \\ -c_m & c_R + c_m \end{bmatrix} \frac{F}{m} \qquad (5.2b)$$

and the line voltages and currents are

$$\mathbf{V}(z,t) = \begin{bmatrix} V_G(z,t) \\ V_R(z,t) \end{bmatrix}$$

$$\mathbf{I}(z,t) = \begin{bmatrix} I_G(z,t) \\ I_R(z,t) \end{bmatrix} \qquad (5.3)$$

These can be expanded as

$$\frac{\partial V_G(z,t)}{\partial z} = -l_G \frac{\partial I_G(z,t)}{\partial t} - l_m \frac{\partial I_R(z,t)}{\partial t}$$

$$\frac{\partial V_R(z,t)}{\partial z} = -l_m \frac{\partial I_G(z,t)}{\partial t} - l_R \frac{\partial I_R(z,t)}{\partial t} \qquad (5.4a)$$

and

$$\frac{\partial I_G(z,t)}{\partial z} = -(c_G + c_m) \frac{\partial V_G(z,t)}{\partial t} + c_m \frac{\partial V_R(z,t)}{\partial t}$$

$$\frac{\partial I_R(z,t)}{\partial z} = c_m \frac{\partial V_G(z,t)}{\partial t} - (c_R + c_m) \frac{\partial V_R(z,t)}{\partial t} \qquad (5.4b)$$

The sinusoidal steady-state (phasor) solution is obtained by exciting the driven line with a single-frequency sinusoidal source, $V_S(t) = V_S\cos(\omega t + \theta_S)$, giving similar forms of the line voltage and current: $V(z,t) = V\cos(\omega t + \theta_V)$ and $I(z,t) = I\cos(\omega t + \theta_I)$. These are symbolized as $\hat{V}_S = V_S\angle\theta_S$, $\hat{V} = V\angle\theta_V$, and $\hat{I} = I\angle\theta_I$. The MTL equations are obtained by replacing all derivatives with respect to time t as $\partial/\partial t \Rightarrow j\omega$:

$$\frac{d}{dz}\hat{\mathbf{V}}(z) = -j\omega\mathbf{L}\hat{\mathbf{I}}(z)$$

$$\frac{d}{dz}\hat{\mathbf{I}}(z) = -j\omega\mathbf{C}\hat{\mathbf{V}}(z) \qquad (5.5)$$

and the MTL equations become ordinary differential equations independent of time, t. These phasor equations can be expanded similarly as

$$\frac{d\hat{V}_G(z)}{dz} = -j\omega l_G \hat{I}_G(z) - j\omega l_m \hat{I}_R(z)$$

$$\frac{d\hat{V}_R(z)}{dz} = -j\omega l_m \hat{I}_G(z) - j\omega l_R \hat{I}_R(z)$$

(5.6a)

and

$$\frac{d\hat{I}_G(z)}{dz} = -j\omega(c_G + c_m)\hat{V}_G(z) + j\omega c_m \hat{V}_R(z)$$

$$\frac{d\hat{I}_R(z)}{dz} = j\omega c_m \hat{V}_G(z) - j\omega(c_R + c_m)\hat{V}_R(z)$$

(5.6b)

These expanded equations can be written in a more illustrative form by moving the self-inductance and self-capacitance terms to the left-hand sides of the equations. Leaving the mutual terms remaining on the right-hand sides of the equations allows them to be viewed as voltage and current sources due to the other line. Finally, we group these into equations for the generator line and for the receptor line as

$$\frac{\partial V_G(z, t)}{\partial z} + l_G \frac{\partial I_G(z, t)}{\partial t} = -l_m \frac{\partial I_R(z, t)}{\partial t}$$

$$\frac{\partial I_G(z, t)}{\partial z} + (c_G + c_m) \frac{\partial V_G(z, t)}{\partial t} = c_m \frac{\partial V_R(z, t)}{\partial t}$$

(5.7a)

and

$$\frac{\partial V_R(z, t)}{\partial z} + l_R \frac{\partial I_R(z, t)}{\partial t} = -l_m \frac{\partial I_G(z, t)}{\partial t}$$

$$\frac{\partial I_R(z, t)}{\partial z} + (c_R + c_m) \frac{\partial V_R(z, t)}{\partial t} = c_m \frac{\partial V_G(z, t)}{\partial t}$$

(5.7b)

The important observation here is that

(5,7) shows that the coupled lines can be viewed as two UNCOUPLED two-conductor lines with sources on the right-hand sides of the equations representing the mutual inductances and mutual capacitances: voltage sources representing

$$-l_m \frac{\partial I_R(z,t)}{\partial t} \quad \text{and} \quad -l_m \frac{\partial I_G(z,t)}{\partial t}$$

and current sources representing

$$c_m \frac{\partial V_R(z,t)}{\partial t} \quad \text{and} \quad c_m \frac{\partial V_G(z,t)}{\partial t}$$

If we consider *electrically short lines*, $\mathscr{L} < \frac{1}{10}\lambda$, we can construct the lumped-circuit model shown in Fig. 5.1. The *total* line inductances and capacitances are the per-unit-length parameters multiplied by the line length and give the total parameters as $L_m = l_m\mathscr{L}$, $L_G = l_G\mathscr{L}$, $L_R = l_R\mathscr{L}$, $C_m = c_m\mathscr{L}$, $C_G + C_m = (c_G + c_m)\mathscr{L}$, and $C_R + C_m = (c_R + c_m)\mathscr{L}$.

To simplify the model further, we ignore the self-inductances and self-capacitances, giving the reduced equivalent circuit shown in Fig. 5.2.

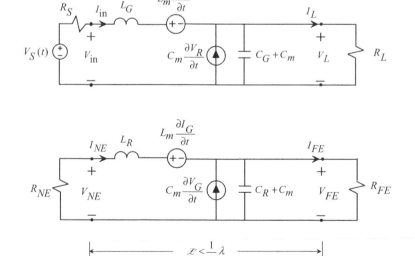

FIGURE 5.1. Coupled MTL equations with an alternative view for *electrically short* lines.

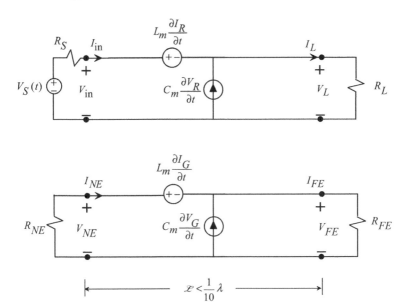

FIGURE 5.2. Simplified model neglecting the self-inductances and self-capacitances.

5.1 THE INDUCTIVE–CAPACITIVE COUPLING APPROXIMATE MODEL

Finally, we will obtain a very simple, but *approximate*, crosstalk prediction model. There are two restrictions on its use:

1. *The lines must be weakly coupled.* The voltage and current in the driven or generator circuit, V_G and I_G, create electric and magnetic fields that couple to the receptor circuit and induce voltage V_R and current I_R in it. In turn, the voltage V_R and current I_R also generate electric and magnetic fields that couple back to the generator circuit, producing additional voltages and currents that add to V_G and I_G, and so on. The lines are said to be *weakly coupled* if the voltage and current induced in the receptor circuit, V_R and I_R, are much smaller than V_G and I_G, so that the additional voltage and current contributions induced back into the generator circuit from the receptor circuit are insignificant and can be ignored. Therefore, the voltage and current sources in the generator line: $L_m(\partial I_R/\partial t)$ and $C_m(\partial V_R/\partial t)$ can be ignored, leaving only the sources in the receptor line: $L_m(\partial I_G/\partial t)$ and $C_m(\partial V_G/\partial t)$. Hence for weakly coupled lines, this induction is *approximately* a one-way process: from the generator circuit to the receptor circuit. We can quantitatively judge whether this is satisfied by examining the inductive

coupling coefficient between the two lines:

$$k = \frac{l_m}{\sqrt{l_G l_R}} \tag{5.8}$$

This is like the coupling coefficient between the primary and the secondary of a transformer. If $k \ll 1$, we say that the circuits are weakly coupled.

2. *The second criterion is that the line must be electrically very short at the largest significant frequency of $V_S(t)$(its bandwidth).* Essentially, this means that *the voltage and current of the generator line are essentially the same at all points along the generator line (independent of z),* so we can say that, approximately,

$$V_G(t) \cong \frac{R_L}{R_S + R_L} V_S(t) \qquad \text{for all } 0 \le z \le \mathscr{L}$$

$$I_G(t) \cong \frac{V_S(t)}{R_S + R_L} \qquad \text{for all } 0 \le z \le \mathscr{L}$$

Hence the voltage and current of the generator line are as though they were dc.

For a line of total length, $\mathscr{L}$ we may represent the coupled lines approximately as shown in Fig. 5.3, and the total *mutual* inductance and *mutual* capacitance are denoted as $L_m = l_m \mathscr{L}$ and $C_m = c_m \mathscr{L}$. We will neglect the self-inductances, l_G and l_R, as well as the self-capacitances, c_G and c_R.

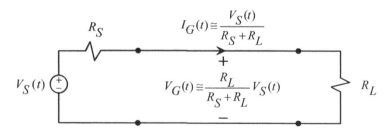

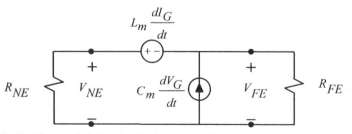

FIGURE 5.3. Weakly coupled electrically short crosstalk prediction approximate model.

Since the line is electrically short, the line voltages and currents are independent of position along it and the partial derivatives become ordinary derivatives: $\partial/\partial t \Rightarrow d/dt$. From this approximate circuit, the crosstalk voltages are obtained, by superposition, as

$$V_{NE}(t) \cong \frac{R_{NE}}{R_{NE} + R_{FE}} L_m \frac{dI_G}{dt} + \frac{R_{NE}R_{FE}}{R_{NE} + R_{FE}} C_m \frac{dV_G}{dt} \tag{5.9a}$$

$$V_{FE}(t) \cong -\frac{R_{FE}}{R_{NE} + R_{FE}} L_m \frac{dI_G}{dt} + \frac{R_{NE}R_{FE}}{R_{NE} + R_{FE}} C_m \frac{dV_G}{dt} \tag{5.9b}$$

Substituting the approximate relations for V_G and I_G gives

$$\boxed{V_{NE}(t) \cong \left(\frac{R_{NE}}{R_{NE} + R_{FE}} L_m + \frac{R_{NE}R_{FE}}{R_{NE} + R_{FE}} C_m R_L \right) \frac{1}{R_S + R_L} \frac{dV_S(t)}{dt}} \tag{5.10a}$$

$$\boxed{V_{FE}(t) \cong \left(-\frac{R_{FE}}{R_{NE} + R_{FE}} L_m + \frac{R_{NE}R_{FE}}{R_{NE} + R_{FE}} C_m R_L \right) \frac{1}{R_S + R_L} \frac{dV_S(t)}{dt}} \tag{5.10b}$$

This shows that the near and far-end crosstalk voltages are proportional to the derivative of $V_S(t)$! Hence, if the source voltage $V_S(t)$ is a trapezoidal digital waveform having rise time τ_r and fall time τ_f and the restrictions on this model apply, the near and far-end time-domain crosstalk voltage waveforms should appear to be rectangular pulses occurring during the rise and fall times of the pulse train, and the heights of the pulses should be proportional to the *slopes* of $V_S(t)$: A/τ_r and A/τ_f.

The frequency-domain, phasor crosstalk voltages are obtained by substituting $j\omega$ for all time derivatives:

$$\boxed{\hat{V}_{NE} = V_{NE} \angle \theta_{NE} \cong j\omega \left(\frac{R_{NE}}{R_{NE} + R_{FE}} L_m + \frac{R_{NE}R_{FE}}{R_{NE} + R_{FE}} C_m R_L \right) \frac{1}{R_S + R_L} \hat{V}_S}$$

$$\tag{5.11a}$$

$$\boxed{\hat{V}_{FE} = V_{FE} \angle \theta_{FE} \cong j\omega \left(-\frac{R_{FE}}{R_{NE} + R_{FE}} L_m + \frac{R_{NE}R_{FE}}{R_{NE} + R_{FE}} C_m R_L \right) \frac{1}{R_S + R_L} \hat{V}_S}$$

$$\tag{5.11b}$$

This shows that for resistive terminations the frequency-response crosstalk voltage *transfer functions*, $\hat{V}_{NE}/\hat{V}_S$ and $\hat{V}_{FE}/\hat{V}_S$, increase linearly with frequency and have a phase angle of $\pm 90°$.

This model is said to be the *inductive-capacitive coupling model* since the crosstalk is produced approximately by the total mutual inductance between the two circuits, L_m, and the total mutual capacitance between the two circuits, C_m. The self-inductances and self-capacitances are ignored. Equations (5.11) show that for the frequencies where the line is electrically short and has resistive terminations, the frequency-domain crosstalk voltage transfer functions must increase linearly with frequency. Hence we can obtain a criterion for when the line is electrically short for a particular time-domain source voltage waveform, $V_S(t)$, and when this model is valid. If $V_S(t)$ represents a digital clock waveform with equal rise and fall times, the bandwidth is approximately BW $\cong 1/\tau_r$. For the line to be electrically very short at this maximum frequency, we must have

$$\mathscr{L} \ll \frac{1}{10} \frac{v}{\text{BW}}$$

or

$$\boxed{\tau_r \gg 10 T_D} \qquad (5.12)$$

EXAMPLE

As an example, consider the PCB configuration shown in Fig. 5.4.

We can calculate an approximate effective relative permittivity, $\varepsilon'_r = (1 + 4.7)/2 = 2.85$, as the average of the relative permittivities of air and glass epoxy (FR4). Hence the velocities of propagation are on the order of 60% of the speed of light, and the one-way time delay is on the order of $T_D \cong 1.4$ ns. For this inductive–capacitive approximate model to be valid, the rise and fall times of the source should be much greater than about 14 ns. Hence for the lines to be electrically very short, the rise and fall times should be greater than perhaps 100 ns. So we will arbitrarily set the rise and fall times of the pulse to 100 ns and the period to 500 ns (2 MHz). The coupling coefficient is $k = l_m/\sqrt{l_G l_R} = 0.559$. This is bordering on the boundary between weakly coupled and strongly coupled. The exact per-unit-length inductance and capacitance matrices are computed numerically with the PCB. FOR computer

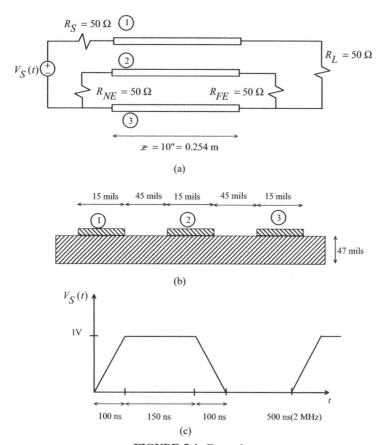

FIGURE 5.4. Example.

program described in Chapter 4 as

$$\mathbf{L} = \begin{bmatrix} 1.383 & 0.691 \\ 0.691 & 1.107 \end{bmatrix} \frac{\mu\mathrm{H}}{\mathrm{m}}$$

$$\mathbf{C} = \begin{bmatrix} 29.69 & -20.26 \\ -20.26 & 40.52 \end{bmatrix} \frac{\mathrm{pF}}{\mathrm{m}}$$

The exact time-domain near- and far-end crosstalk voltages computed using the PSPICE subcircuit model described in Chapter 6 are shown in Fig. 5.5. Observe that the crosstalk voltages appear, as expected, as rectangles during the rise and fall times of $V_S(t)$, which are $0 \le t \le 100\,\mathrm{ns}$ and $250 \le t \le 350\,\mathrm{ns}$. From numerical computations in Chapter 4, $l_m = 6.9157 \times 10^{-7}$ and $c_m = 2.02619 \times 10^{-11}$. Since the total line length is $\mathscr{L} = 0.254\,\mathrm{m}$, the total mutual inductance and capacitance are

$L_m = l_m \mathscr{L} = 1.757 \times 10^{-7}$ H   and $C_m = c_m \mathscr{L} = 5.147 \times 10^{-12}$ F.   The derivative (slew rate) of $V_S(t)$ is

$$\frac{dV_S(t)}{dt} = \frac{1 V}{100 \text{ ns}} = 10^7 \frac{V}{s}$$

Hence from the inductive-coupling model we compute the near- and far-end crosstalk voltages [the heights of the pulses occurring during the rise and fall times of $V_S(t)$] from (5.10a) as

$$V_{NE}(t) \cong \left(\frac{R_{NE}}{R_{NE} + R_{FE}} L_m + \frac{R_{NE}R_{FE}}{R_{NE} + R_{FE}} C_m R_L \right) \frac{1}{R_S + R_L} \frac{dV_S(t)}{dt}$$

$$= 9.426 \text{ mV}$$

$$V_{FE}(t) \cong \left(-\frac{R_{FE}}{R_{NE} + R_{FE}} L_m + \frac{R_{NE}R_{FE}}{R_{NE} + R_{FE}} C_m R_L \right) \frac{1}{R_S + R_L} \frac{dV_S(t)}{dt}$$

$$= -8.14 \text{ mV}$$

The *exact* voltages computed using a PSPICE subcircuit model developed in Chapter 6 are shown in Fig. 5.5 and are $V_{NE} = 9.4263$ mV and $V_{FE} = -8.1396$ mV!
   The previous calculations investigated the *time-domain* crosstalk where the source voltage, $V_S(t)$, was a trapezoidal digital waveform. We next investigate the *frequency-domain* response of the crosstalk to obtain further information into the causes of the time-domain crosstalk. To sketch the frequency response, we drive the line with a single-frequency sinusoid, $V_S(t) = V_S \cos(\omega t + \theta_S)$. To accomplish this, we replace all time derivatives with $\partial/\partial t \Rightarrow j\omega$ and perform our calculations with the phasor circuit model. The resulting crosstalk waveforms are also single-frequency sinusoids, $V_{NE}(t) = V_{NE}\cos(\omega t + \theta_{NE})$ and $V_{FE}(t) = V_{FE}\cos(\omega t + \theta_{FE})$, where the task is to determine, given the magnitude and phase of the source, $\hat{V}_S = V_S \angle \theta_S$, the magnitudes and phases of the crosstalk voltages: $\hat{V}_{NE} = V_{NE} \angle \theta_{NE}$ and $\hat{V}_{FE} = V_{FE} \angle \theta_{FE}$. The solutions in (5.11) show that the angles of the crosstalk for resistive terminations are $\pm 90°$. A few preliminary calculations are in order. We will plot the *transfer function* magnitudes, V_{NE}/V_S and V_{FE}/V_S, versus frequency. For the time-domain problem just investigated, we will be interested in identifying the line length in terms of the frequencies where it is one wavelength, $\mathscr{L} = \lambda$, and electrically short, $\mathscr{L} = \frac{1}{10}\lambda$, as well as the bandwidth of the source pulse waveform, $BW = 1/\tau_r$. These are best determined in terms of the line one-way time delay of (approximately) $T_D = \mathscr{L}/v_0\sqrt{\varepsilon'_r} = 1.43$ ns where we have used $\mathscr{L} = 0.254$ m, $v_0 = 3 \times 10^8$ m/s and an approximate, effective permittivity,

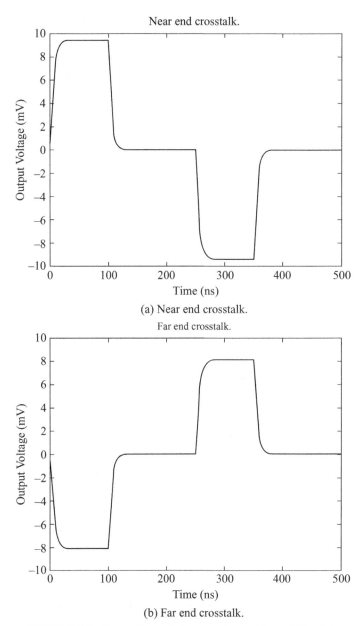

FIGURE 5.5. Crosstalk predictions for the problem of Fig. 5.4.

$\varepsilon'_r = (1 + 4.7)/2 = 2.85$. For a rise and fall time of 100 ns, the BW is BW $= 1/\tau_r = 10$ MHz. The fundamental frequency is $f_0 = 1/T = 1/500$ ns $= 2$ MHz. The line is one wavelength at $f|_{\mathscr{L}=\lambda} = 1/T_D = 700$ MHz and is electrically short at $f|_{\mathscr{L}=(1/10)\lambda} = 1/10T_D = 70$ MHz. Hence the line is

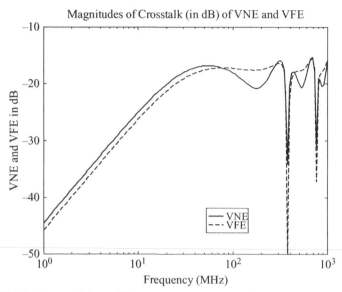

FIGURE 5.6. Near- and far-end frequency-response crosstalk computed with PSPICE.

very short, electrically, for this problem and the approximate inductive–capacitive coupling model is quite adequate, as we have seen.

The frequency responses of the magnitudes of the crosstalk from 1 MHz to 1 GHz are shown in Fig. 5.6 for $\hat{V}_S = 1 \angle 0°$. The line is one wavelength at around 700 MHz and is electrically short at around 70 MHz. The spectrum of the digital pulse train in Fig. 5.4 occupies from $f_0 = 1/T = 1/500\,\text{ns} = 2\,\text{MHz}$ to $\text{BW} = 1/\tau_r = 10\,\text{MHz}$, where the line is clearly very short electrically, and the transfer function magnitudes clearly increase with frequency as 20 dB/ decade, which is implied by the frequency-domain solution in (5.11). Observe in the frequency plot in Fig. 5.6 that the magnitudes of V_{NE} and V_{FE} are not the same at the low frequencies even though all of the terminal resistances are the same: 50 Ω. This difference is due to the minus sign in the inductive coupling portion of V_{FE} in (5.11b) that is due to L_m. The inductive coupling due to L_m and the capacitive coupling due to C_m subtract in V_{FE}, whereas they add in V_{NE} in (5.11a).

5.2 SEPARATION OF THE CROSSTALK INTO INDUCTIVE AND CAPACITIVE COUPLING COMPONENTS

The crosstalk solutions are the sum of two components. One component is due to the total mutual inductance, L_m, and is referred to as the *inductive coupling* between the two circuits. The other is due to the total mutual capacitance, C_m, and is referred to as the *capacitive coupling* between the two circuits. This is emphasized for both the time- and frequency-domain solutions. The time-domain solution is

$$V_{NE}(t) \cong \left[\underbrace{\frac{R_{NE}}{(R_{NE} + R_{FE})(R_S + R_L)} L_m}_{\text{inductive coupling}} + \underbrace{\frac{R_{NE}R_{FE}R_L}{(R_{NE} + R_{FE})(R_S + R_L)} C_m}_{\text{capacitive coupling}} \right] \frac{dV_S(t)}{dt}$$

(5.13a)

$$V_{FE}(t) \cong \left[-\underbrace{\frac{R_{FE}}{(R_{NE} + R_{FE})(R_S + R_L)} L_m}_{\text{inductive coupling}} + \underbrace{\frac{R_{NE}R_{FE}R_L}{(R_{NE} + R_{FE})(R_S + R_L)} C_m}_{\text{capacitive coupling}} \right] \frac{dV_S(t)}{dt}$$

(5.13b)

and the frequency-domain solution is

$$\hat{V}_{NE} = V_{NE} \angle \theta_{NE} \cong j\omega \left[\underbrace{\frac{R_{NE}}{(R_{NE} + R_{FE})(R_S + R_L)} L_m}_{\text{inductive coupling}} + \underbrace{\frac{R_{NE}R_{FE}R_L}{(R_{NE} + R_{FE})(R_S + R_L)} C_m}_{\text{capacitive coupling}} \right] \hat{V}_S$$

(5.14a)

$$\hat{V}_{FE} = V_{FE} \angle \theta_{FE} \cong j\omega \left[-\underbrace{\frac{R_{FE}}{(R_{NE} + R_{FE})(R_S + R_L)} L_m}_{\text{inductive coupling}} + \underbrace{\frac{R_{NE}R_{FE}R_L}{(R_{NE} + R_{FE})(R_S + R_L)} C_m}_{\text{capacitive coupling}} \right] \hat{V}_S$$

(5.14b)

In the example, all the termination resistances are $50\,\Omega$, and the results reduce to

$$V_{NE}(t) \cong \left[\underbrace{8.79 \times 10^{-10}}_{\text{inductive coupling}} + \underbrace{0.643 \times 10^{-10}}_{\text{capacitive coupling}} \right] \frac{dV_S(t)}{dt}$$

$$V_{FE}(t) \cong \left[\underbrace{-8.79 \times 10^{-10}}_{\text{inductive coupling}} + \underbrace{0.643 \times 10^{-10}}_{\text{capacitive coupling}} \right] \frac{dV_S(t)}{dt}$$

and

$$V_{NE}\angle\theta_{NE} \cong j\omega \left[\underbrace{8.79 \times 10^{-10}}_{\text{inductive coupling}} + \underbrace{0.643 \times 10^{-10}}_{\text{capacitive coupling}} \right] \hat{V}_S$$

$$V_{FE}\angle\theta_{FE} \cong j\omega \left[\underbrace{-8.79 \times 10^{-10}}_{\text{inductive coupling}} + \underbrace{0.643 \times 10^{-10}}_{\text{capacitive coupling}} \right] \hat{V}_S$$

Observe that for this problem the inductive coupling dominates the capacitive coupling by a factor of 13.7. This brings up the question of whether we can expect this intuitively. We expect that inductive coupling would dominate capacitive coupling for "large" currents since inductance depends on current. "Low" termination impedances tend to draw "large" line currents. On the other hand, we expect that capacitive coupling would dominate inductive coupling for "large" voltages and relatively "small" currents on the line, so this makes sense for "high" terminations impedances.

We can quantify this intuition by observing in (5.10) and (5.11) that inductive coupling dominates capacitive coupling if

$$\frac{L_m}{C_m} \gg R_{FE}R_L \qquad \text{in} \quad V_{NE} \qquad\qquad (5.15\text{a})$$

$$\frac{L_m}{C_m} \gg R_{NE}R_L \qquad \text{in} \quad V_{FE} \qquad\qquad (5.15\text{b})$$

and capacitive coupling dominates inductive coupling if

$$\frac{L_m}{C_m} \ll R_{FE}R_L \quad \text{in} \quad V_{NE} \tag{5.16a}$$

$$\frac{L_m}{C_m} \ll R_{NE}R_L \quad \text{in} \quad V_{FE} \tag{5.16b}$$

Hence we obtain the idea that *inductive coupling dominates capacitive coupling for "low"-impedance terminations, and capacitive coupling dominates inductive coupling for "high"-impedance terminations,*

We can further quantify this notion if we define the "characteristic impedances" of the generator and receptor circuits as though they are isolated two-conductor lines. However, each of these circuits is not, strictly speaking, isolated but is approximately so for weak coupling. Recall the relation between the inductance and capacitance matrices for a homogeneous medium in (4.9):

$$\mathbf{LC} = \begin{bmatrix} l_G & l_m \\ l_m & l_R \end{bmatrix} \begin{bmatrix} c_G + c_m & -c_m \\ -c_m & c_R + c_m \end{bmatrix} = \frac{1}{v^2} \begin{bmatrix} 1 & 0 \\ 0 & 1 \end{bmatrix}$$

Multiplying the off-diagonal terms gives

$$\frac{L_m}{C_m} = \frac{l_G}{c_R + c_m}$$

$$\frac{L_m}{C_m} = \frac{l_R}{c_G + c_m}$$

Multiplying these gives

$$\left(\frac{L_m}{C_m}\right)^2 = \frac{l_G}{c_R + c_m} \frac{l_R}{c_G + c_m} \tag{5.17}$$

It seems logical to define the "characteristic impedances" of the generator and receptor circuits as

$$Z_{CG} = \sqrt{\frac{l_G}{c_G + c_m}} \tag{5.18a}$$

$$Z_{CR} = \sqrt{\frac{l_R}{c_R + c_m}} \tag{5.18b}$$

Hence

$$\boxed{\frac{L_m}{C_m} = Z_{CG}Z_{CR}} \tag{5.19}$$

Hence we can quantify this by observing that inductive coupling dominates capacitive coupling if

$$Z_{CG}Z_{CR} \gg R_{FE}R_L \quad \text{in} \quad V_{NE} \tag{5.15a}$$

$$Z_{CG}Z_{CR} \gg R_{NE}R_L \quad \text{in} \quad V_{FE} \tag{5.15b}$$

and capacitive coupling dominates inductive coupling if

$$Z_{CG}Z_{CR} \ll R_{FE}R_L \quad \text{in} \quad V_{NE} \tag{5.16a}$$

$$Z_{CG}Z_{CR} \ll R_{NE}R_L \quad \text{in} \quad V_{FE} \tag{5.16b}$$

Hence we obtain the idea that *inductive coupling dominates capacitive coupling for "low"-impedance terminations, and capacitive coupling dominates inductive coupling for "high"-impedance terminations.* The terms "high impedance" and "low impedance" are used in the sense "with respect to the characteristic impedances of the circuits: Z_{CG} and Z_{CR}."

We will find these concepts to be very useful in explaining when and why shielded wires are useful in reducing crosstalk in "high-impedance" circuits, whereas twisted pairs of wires are useful in reducing crosstalk in "low-impedance" circuits. Shielded wires generally affect only the capacitive component of crosstalk, whereas twisted pairs of wires generally affect only the inductive component of crosstalk. This explains why you tend to see shielded wires frequently used to reduce crosstalk in CMOS (high-impedance) circuits, and twisted pairs of wires frequently used in reducing crosstalk in power supplies (low impedance circuits)! This is the key to understanding when and why shielded wires may OR MAY NOT reduce crosstalk and when and why twisted pairs of wires may OR MAY NOT reduce crosstalk! This separation of the total crosstalk into the sum of an inductive coupling component, and a capacitive coupling component is the key to dealing effectively with crosstalk. You will talk to many EMC engineers who say either "I use shielded wires to reduce crosstalk; twisted pairs of wires never work for me" or "I use twisted pairs of wires to reduce crosstalk; shielded wires never work for me." BOTH ARE CORRECT in their own worlds. The first engineer works exclusively in CMOS design (high impedance, where capacitive crosstalk dominates), whereas the other one works exclusively in power supply design (low impedance, where inductive crosstalk dominates). You will find in your everyday work that this separation of the total crosstalk into the sum of an inductive coupling component and a capacitive coupling component completely removes all previous mysteries that you have been struggling to resolve and understand.

Figure 5.7 gives a graphical explanation of these concepts in terms of the frequency-domain transfer functions. The lines are electrically short where the crosstalk increases linearly with frequency at a rate of 20 dB/decade. If you want to reduce the TOTAL crosstalk in a low-impedance circuit, Fig. 5.7(a) shows that you MUST reduce the *dominant* inductive coupling

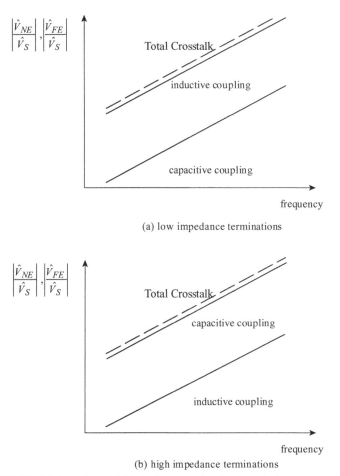

FIGURE 5.7. Explaining when shielded wires or twisted pairs of wires may or MAY NOT reduce crosstalk, depending on the termination impedance levels of the circuit.

component. Reducing the capacitive coupling will have little or NO EFFECT on the total crosstalk since it is dominated by the inductive coupling component. Eliminating the inductive component will drop the crosstalk to the underlying capacitive component of the crosstalk. Similarly, if you want to reduce the TOTAL crosstalk in a high-impedance circuit, Fig. 5.7(b) shows that you MUST reduce the *dominant* capacitive coupling component. Reducing the inductive coupling will have little or NO EFFECT on the total crosstalk since it is dominated by the capacitive coupling component. Eliminating the capacitive component will drop the crosstalk to the underlying inductive component of the crosstalk.

5.3 COMMON-IMPEDANCE COUPLING

One final coupling mechanism that is very easy to understand and which dominates the crosstalk at low frequencies is *common-impedance coupling*. This is caused by the impedance of the reference conductor, R_0, and is shown in Fig. 5.8. Since the impedance (resistance) of the reference conductor is

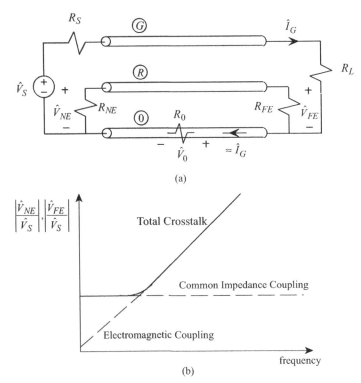

FIGURE 5.8. Common-impedance coupling.

much smaller than any of the termination impedances, the majority of the generator line current, I_G, returns through it. This gives, at low frequency, an ohmic voltage developed between two ends of it of

$$\hat{V}_0 \cong R_0 \hat{I}_G \tag{5.20}$$

where R_0 is the ohmic resistance of the reference conductor. This is voltage-divided between the two terminal impedances of the receptor circuit to give a component of the crosstalk of

$$\hat{V}_{NE}^{CI} = \frac{R_{NE}}{R_{NE} + R_{FE}} R_0 \frac{\hat{V}_S}{R_S + R_L} \tag{5.21a}$$

$$\hat{V}_{FE}^{CI} = -\frac{R_{FE}}{R_{NE} + R_{FE}} R_0 \frac{\hat{V}_S}{R_S + R_L} \tag{5.21b}$$

This tends to dominate the crosstalk at low frequencies where the electromagnetic inductive and capacitive crosstalk decreases at a rate of 20 dB/decade (linearly with frequency) when the electromagnetic inductive and capacitive crosstalk component falls below the common impedance coupling.

5.4   EFFECT OF SHIELDED WIRES IN REDUCING CROSSTALK

In this section we investigate the use of shielded wires in reducing crosstalk. Figure 5.9 shows the usual configuration of a generator wire and a receptor wire above a ground plane. The inductive and capacitive coupling is such that one dominates the other, depending on the line terminal impedances. We have added a circular cylindrical shield around and centered on the receptor wire in order to eliminate or reduce the crosstalk to the receptor circuit. We have shown switches that are used to connect the ends of the shield to the ground plane ("grounding it"). To clarify this we will use the symbols OS, SO, and SS. The symbol OS symbolizes that the left end of the shield is open, not "grounded," but the right end is connected to the ground plane or shorted to the ground plane, or "grounded." Similarly, SO symbolizes that the left end of the shield is shorted to the ground plane and the right end is unconnected, and SS symbolizes that both ends of the shield are connected to the ground plane or shorted to it.

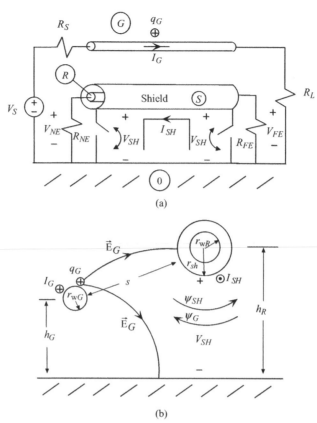

FIGURE 5.9. Crosstalk reduction (time domain or frequency domain) with a shielded wire.

We have arbitrarily placed the shield around the receptor circuit wire. The principle of reciprocity provides that we could alternatively have placed the shield around the generator wire and would have obtained the same result.

A few important observations are in order. There are three *mutual inductance* loops in the problem. Figure 5.10 shows the equivalent receptor-shield circuit. Two loops involve the generator current I_G. One is between the generator current I_G loop and the receptor current I_R loop through the mutual inductance L_{GR}, and a second loop is between the generator current I_G loop and the shield–ground plane current I_{SH} loop through the mutual inductance L_{GS} If the shield is NOT grounded at BOTH ENDS, there can be no path for the shield current and it is zero for SO or OS, so that $I_{SH} = 0$. In order to have a shield current, BOTH ENDS OF THE SHIELD MUST BE GROUNDED: SS. This shield current is limited only by the resistance of the

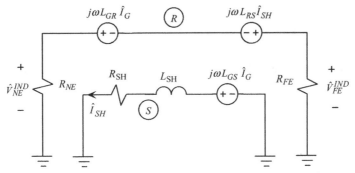

FIGURE 5.10. Equivalent circuit for the shielded receptor wire.

shield, R_{SH}, and its self-inductance, L_{SH}. Hence the induced shield current is determined from

$$\hat{I}_{SH} = \frac{j\omega L_{GS}}{R_{SH} + j\omega L_{SH}} \hat{I}_G \qquad (5.22)$$

This induced shield current, $\hat{I}_{SH}$, induces as a secondary effect, another source in the receptor–ground plane loop through the mutual inductance L_{RS} between the receptor and shield loops, which is the third mutual inductance loop, as shown in Fig. 5.10. The induced shield current $\hat{I}_{SH}$ produces a magnetic flux ψ_{SH}, as shown in Fig. 5.9(b), that tends to cancel the original crosstalk flux from the generator line, ψ_G. The net inductive crosstalk voltages are

$$
\begin{aligned}
\hat{V}_{NE}^{IND} &= \frac{R_{NE}}{R_{NE} + R_{FE}} (j\omega L_{GR}\hat{I}_G - j\omega L_{RS}\hat{I}_{SH}) \\
&= \frac{R_{NE}}{R_{NE} + R_{FE}} j\omega L_{GR}\hat{I}_G \left(1 - \frac{L_{RS}\hat{I}_{SH}}{L_{GR}\hat{I}_G} \right)
\end{aligned}
\qquad (5.23a)
$$

$$
\begin{aligned}
\hat{V}_{FE}^{IND} &= -\frac{R_{FE}}{R_{NE} + R_{FE}} (j\omega L_{GR}\hat{I}_G - j\omega L_{RS}\hat{I}_{SH}) \\
&= -\frac{R_{FE}}{R_{NE} + R_{FE}} j\omega L_{GR}\hat{I}_G \left(1 - \frac{L_{RS}\hat{I}_{SH}}{L_{GR}\hat{I}_G} \right)
\end{aligned}
\qquad (5.23b)
$$

In order to obtain the final result in terms of only the generator current, $\hat{I}_G$, we substitute (5.22), giving

$$\hat{V}_{NE}^{IND} = \frac{R_{NE}}{R_{NE} + R_{FE}} j\omega L_{GR}\hat{I}_G \left[1 - \frac{L_{RS} j\omega L_{GS}}{L_{GR}(R_{SH} + j\omega L_{SH})} \right] \qquad (5.24a)$$

$$\hat{V}_{FE}^{IND} = -\frac{R_{FE}}{R_{NE} + R_{FE}} j\omega L_{GR} \hat{I}_G \left[1 - \frac{L_{RS} j\omega L_{GS}}{L_{GR}(R_{SH} + j\omega L_{SH})} \right] \qquad (5.24b)$$

Some thought will reveal that we have the following relations between the various mutual inductances:

$$L_{GR} = L_{GS} \qquad (5.25a)$$

$$L_{RS} = L_{SH} \qquad (5.25b)$$

The first identity provides that the mutual inductance between the generator–receptor loops, L_{GR}, equals the mutual inductance between the generator–shield loops, L_{GS}. This is rather obvious since the receptor wire and the shield that surrounds it are collocated. The second identity provides that the mutual inductance between the receptor and the shield–ground plane loops, L_{RS}, and the *self-inductance* of the shield–ground plane loop, L_{SH}, are identical. This becomes more intuitive when you recall that the definition of inductance is obtained by applying a current to a loop and determining the magnetic flux through the loop. So putting a current on the shield gives a magnetic flux through the shield–ground plane loop and L_{SH}. Since the receptor wire and the shield are collocated, we obtain the result: $L_{SH} = L_{RS}$.

Hence the result for the inductive crosstalk in (5.24) reduces to

$$\hat{V}_{NE}^{IND} = \frac{R_{NE}}{(R_{NE} + R_{FE})} j\omega L_{GR} \hat{I}_G \left[\frac{R_{SH}}{R_{SH} + j\omega L_{SH}} \right] \qquad (5.26a)$$

$$\underbrace{\qquad\qquad\qquad\qquad}_{\text{crosstalk with the shield removed}} \quad \underbrace{\qquad\qquad}_{\text{effect of shield}}$$

$$\hat{V}_{FE}^{IND} = -\frac{R_{FE}}{R_{NE} + R_{FE}} j\omega L_{GR} \hat{I}_G \left[\frac{R_{SH}}{R_{SH} + j\omega L_{SH}} \right] \qquad (5.26b)$$

$$\underbrace{\qquad\qquad\qquad\qquad}_{\text{crosstalk with the shield removed}} \quad \underbrace{\qquad\qquad}_{\text{effect of shield}}$$

Substituting $\hat{V}_S = (R_S + R_L)\hat{I}_G$ gives

$$\frac{\hat{V}_{NE}^{IND}}{\hat{V}_S} = \frac{R_{NE}}{(R_{NE} + R_{FE})(R_S + R_L)} j\omega L_{GR} \left[\frac{R_{SH}}{R_{SH} + j\omega L_{SH}} \right] \qquad (5.27a)$$

$$\underbrace{\qquad\qquad\qquad\qquad}_{\text{crosstalk with the shield removed}} \quad \underbrace{\qquad\qquad}_{\text{effect of shield}}$$

$$\frac{\hat{V}_{FE}^{IND}}{\hat{V}_S} = -\underbrace{\frac{R_{FE}}{(R_{NE}+R_{FE})(R_S+R_L)}j\omega L_{GR}}_{\text{crosstalk with the shield removed}} \underbrace{\left[\frac{R_{SH}}{R_{SH}+j\omega L_{SH}}\right]}_{\text{effect of shield}} \qquad (5.27b)$$

Hence the inductive crosstalk is simply the product of the crosstalk with the shield removed and a "shielding factor,"

$$\begin{aligned} SF &= \frac{R_{SH}}{R_{SH}+j\omega L_{SH}} \\ &= \frac{1}{1+j\omega(L_{SH}/R_{SH})} \end{aligned} \qquad (5.28)$$

This can be written as

$$SF = \frac{1}{1+j(f/f_b)} \qquad (5.29a)$$

where the *shield break frequency* is

$$f_b = \frac{1}{2\pi}\frac{R_{SH}}{L_{SH}} \qquad (5.29b)$$

and the shield–ground plane circuit time constant is L_{SH}/R_{SH}. Figure 5.11 shows this result for the prediction of the inductive crosstalk. Essentially, this can be summarized as follows:

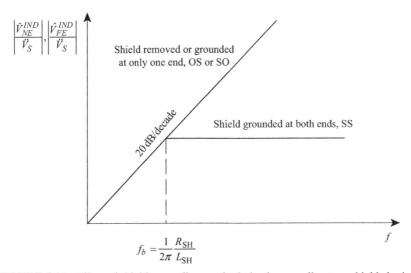

FIGURE 5.11. Effect of shield grounding on the inductive coupling to a shielded wire.

> A shielded wire can reduce inductive coupling ONLY if the shield is grounded at BOTH ends and then only above the shield break frequency.

Above f_b, the generator current returns along the shield since (5.22) shows that above f_b, $\hat{I}_{SH} = (L_{GS}/L_{SH})\hat{I}_G$. Above f_b the shield factor in (5.28) becomes

$$SF = \frac{R_{SH}}{R_{SH} + j\omega L_{SH}}$$

$$\cong \frac{1}{j\omega(L_{SH}/R_{SH})} \quad f > f_b$$

and the $j\omega$ in the numerator and the denominator of (5.27) *cancel.* Hence for $f > f_b$ the transfer function becomes constant independent of frequency as shown in Fig. 5.11:

$$\frac{\hat{V}_{NE}^{IND}}{\hat{V}_S} = \frac{R_{NE}}{R_{NE} + R_{FE}} \frac{R_{SH}L_{GR}}{(R_S + R_L)L_{SH}} \quad f > f_b \qquad (5.30\text{a})$$

$$\frac{\hat{V}_{FE}^{IND}}{\hat{V}_S} = -\frac{R_{FE}}{R_{NE} + R_{FE}} \frac{R_{SH}L_{GR}}{(R_S + R_L)L_{SH}} \quad f > f_b \qquad (5.30\text{b})$$

With regard to a shield affecting the capacitive crosstalk, this is shown in the equivalent circuit in Fig. 5.12 and is summarized as follows:

> A shield that is grounded at EITHER END will eliminate capacitive crosstalk.

We have omitted any self-capacitances. We have also omitted the mutual capacitance between the generator and receptor conductors since the shield surrounds the receptor conductor and forms a Faraday shield so that $C_{GR} = 0$. Grounding either end of the shield (OS, SO, or SS) will, for an electrically short line, render the shield voltage zero, $\hat{V}_{shield} = 0$, at all points along it, thereby eliminating the capacitive coupling in all cases. Another way of looking at this is to recognize that the electric field lines from the generator wire will terminate on the shield and not on the interior receptor wire. In other words, the shield functions as a *Faraday shield* by containing the receptor wire.

These results of using a shield to reduce crosstalk are summarized below and are summarized visually in Fig. 5.13.

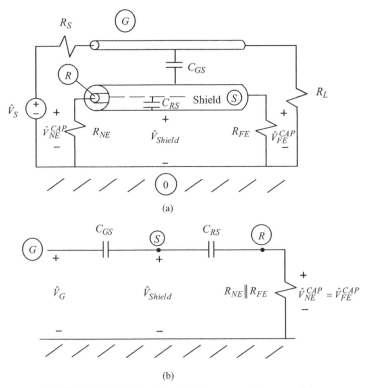

FIGURE 5.12. Effect of a shield on capacitive coupling.

1. *Inductive coupling dominates capacitive coupling (with the shield removed and low-impedance line terminations)* in Fig. 5.13(*a*). To reduce the *total crosstalk,* we must eliminate the *dominant* (inductive) component. Placing the shield around the receptor wire and grounding it at either end (OS or SO) or both ends (SS) will eliminate all the capacitive coupling but will reduce the dominant inductive coupling only above the shield factor frequency of

$$f_b = \frac{1}{2\pi}\frac{R_{SH}}{L_{SH}}$$

   Hence when inductive coupling is dominant as for *low-impedance terminations,* a shield will NOT reduce the *total crosstalk* [except above $f_b = (1/2\pi)/(R_{SH}/L_{SH})$].

2. *Capacitive coupling dominates inductive coupling (with the shield removed and high-impedance line terminations)* in Fig. 5.13(*b*). To reduce the *total crosstalk,* we must eliminate the *dominant* (capacitive) component. Placing the shield around the receptor wire and grounding it at either end (OS or SO) or both ends (SS) will eliminate all the capacitive coupling. Since the capacitive coupling was dominant, the total crosstalk will not be eliminated but will be reduced to the underlying inductive coupling.

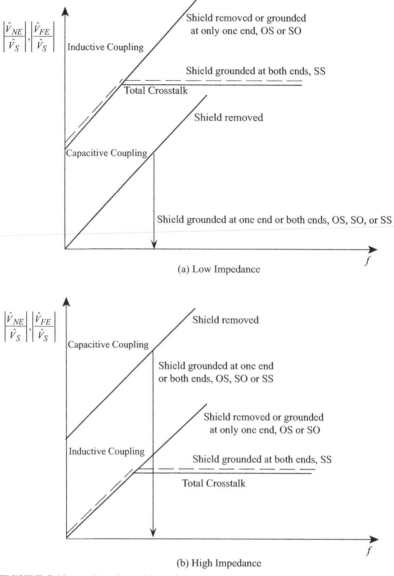

FIGURE 5.13. Explanation of impedance levels on crosstalk reduction by a shield.

Equations for the various inductances and capacitances were obtained using the ideas in Section 4.2.1. Consider dimensions of the line cross sections in Fig. 5.9(b). The shield resistance is

$$R_{SH} = \frac{\mathscr{L}}{\sigma 2\pi r_{sh} t_{sh}} \tag{5.31}$$

where $\sigma = \sigma_{CU} = 5.8 \times 10^7$ is the conductivity of the copper, r_{sh} is the interior radius of the shield, and t_{sh} is the shield thickness. The self-inductance of the shield-ground plane loop is

$$L_{SH} = \frac{\mu_0}{2\pi} \mathscr{L} \ln\left(\frac{2h_R}{r_{sh} + t_{sh}}\right) \tag{5.32}$$

The self-inductances of the generator and receptor circuits (not used) are

$$L_G = \frac{\mu_0}{2\pi} \mathscr{L} \ln\frac{2h_G}{r_{wG}} \tag{5.33}$$

$$L_R = \frac{\mu_0}{2\pi} \mathscr{L} \ln\frac{2h_R}{r_{wR}} \tag{5.34}$$

The mutual inductances between the generator and receptor and between the generator and shield loops are

$$L_{GR} = L_{GS} = \frac{\mu_0}{4\pi} \mathscr{L} \ln\left(1 + 4\frac{h_G h_R}{s^2}\right) \tag{5.35}$$

The inductances of the shield–ground plane loop, and between the receptor and shield loops are

$$L_{SH} = L_{RS} = \frac{\mu_0}{2\pi} \mathscr{L} \ln\frac{2h_R}{r_{sh} + t_{sh}} \tag{5.36}$$

The mutual capacitance between the receptor and the shield is

$$C_{RS} = \frac{2\pi\varepsilon_0\varepsilon_r}{\ln r_{sh}/r_{wR}} \mathscr{L} \tag{5.37}$$

The other capacitances are obtained from

$$\begin{bmatrix} c_G + c_{GS} & -c_{GS} \\ -c_{GS} & c_R + c_{GS} \end{bmatrix} = \mu_0\varepsilon_0 \begin{bmatrix} l_G & l_{GS} \\ l_{GS} & l_R \end{bmatrix}^{-1} \tag{5.38}$$

The entries in (5.38) are per unit length, so any results from it must be multiplied by the shield length, $\mathscr{L}$, to give the total parameter for the line.

5.4.1 Experimental Results

A considerable amount of experimental data in the frequency domain will be shown in this chapter. These data show the experimental effects of shielded wires and twisted wire pairs. The experimental frequency points will be at 1, 1.5, 2, 2.5, 3, 4, 5, 6, 7, 8, and 9 in each decade. The frequency points will be in six decades from 100 Hz to 100 MHz. In all the crosstalk measured in this chapter the near end phasor (frequency–domain) crosstalk, V_{NE}, be measured. In this subsection we show some experimental data that confirm the previous model predictions. The experimental setup used is shown in Fig. 5.14. A shield is placed around the receptor wire and its ends are either not connected to the ground plane, "open," O, or "shorted" to the ground plane, S. The total line length is 3.6576 m or 12 ft. The line is one-tenth of a wavelength and electrically short at 8.2 MHz. The source resistance is zero, $R_S = 0$, and all the other terminations are identical resistors, R. We investigate two terminations: $R = 50\,\Omega$ (inductive coupling is dominant) and $R = 1\,k\Omega$ (capacitive coupling is dominant). Four shield groundings are shown: OO, SO, OS, SS. The generator wire's insulation is touching the insulation around the shield,

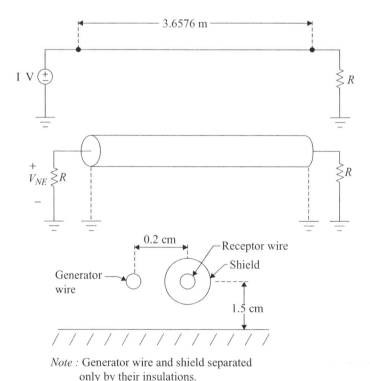

Note : Generator wire and shield separated
only by their insulations.

FIGURE 5.14. Experiment to illustrate the effect of shield grounding on crosstalk to a shielded wire.

giving a separation of approximately 0.2 cm, so that they are as close as possible. The wires are at a height of 1.5 cm above the ground plane. The generator wire is a No 20 solid wire ($r_{wG} = 15$ mils), and the shielded wire are ($r_{wR} = 15$ mils)(No.22   gauge   stranded, 7×30), $\varepsilon_r = 2.1$ (Teflon), $r_{sh} = 35$ mils, $t_{sh} = 5$ mils. The resistance of the shield is 89.9 mΩ. The total inductances   are $L_G = 3.15\,\mu$H, $L_R = 3.19\mu$ H, L_{RS}, $L_{SH} = 2.48\,\mu$H, and $L_{GR} = L_{GS} = 1.98\mu$H. The total capacitances are $C_{RS} = 503.6$ pF and $C_{GS} = 76.3$ pF, and the shield break frequency is $f_{SH} = R_{SH}/2\pi L_{SH} = 5.77$ kHz.

Figure 5.15(a) shows the frequency-domain crosstalk transfer functions from 100 Hz to 100 MHz for $R = 50\,\Omega$. Compare this to Fig. 5.13(a) to see that the inductive coupling is dominant, as expected, since $R = 50\,\Omega$ is less than the characteristic impedance of the lines. All four cases of shield termination are shown: OO, OS, SO, SS. Note that for SS, the inductive coupling flattens out above the shield break frequency, which is

$$f_b = \frac{1}{2\pi} \frac{R_{SH}}{L_{SH}} = 5.8\,\text{kHz}$$

Figure 5.15(b) shows the (dominant) capacitive crosstalk for $R = 1\,\text{k}\Omega$. For OS, SO, and SS the capacitive coupling is removed and the crosstalk is reduced to the inductive coupling. [see Fig. 5.13(b)]. Notice that SO and OS (both inductive) start to diverge at 200 kHz. This is due to our assumption that if you ground one end of the shield the other end will essentially be grounded for an electrically short line. This shows that is true only for an electrically very short section of the line.

In both plots there is a notation of "PIGTAILS $= 0.5$ CM." Pigtails are exposed terminals of shields that we discuss in the next section. The notation of "PIGTAILS $= 0.5$ CM" essentially means that the shields when shorted were connected such that the receptor wire of the shield was not substantially exposed. Figure 5.16 essentially summarizes these results, thereby verifying the model.

5.5   EFFECT OF SHIELD PIGTAILS

Essentially, "pigtails" refer to exposed sections of a shield that exposes the interior wire where the shield is stripped back to make a connection within the connector, as shown in Fig. 5.17. Although this seems to be inconsequential, we will provide experimental data that will show that this can cause serious crosstalk problems. A simple way of seeing this is as follows. Suppose that the shielded wire is very long, say, 3.6576 m, as we use in the experiment. Now

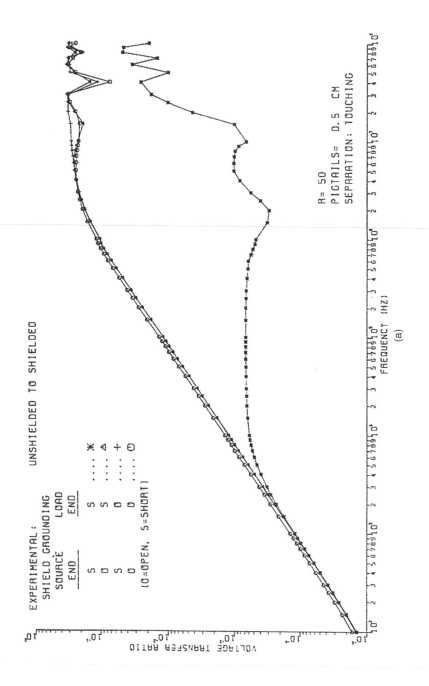

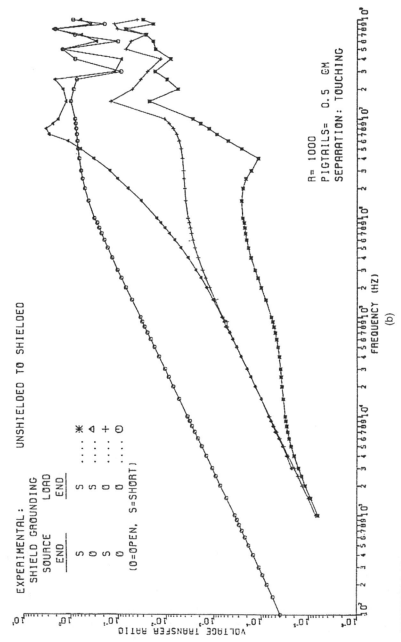

FIGURE 5.15. Measured near-end crosstalk for the configuration of Fig. 5.14 for the shield ungrounded, grounded at the left end, grounded at the right end, and grounded at both ends for (a) $R = 50\,\Omega$ and (b) $R = 1\,\mathrm{k}\Omega$.

185

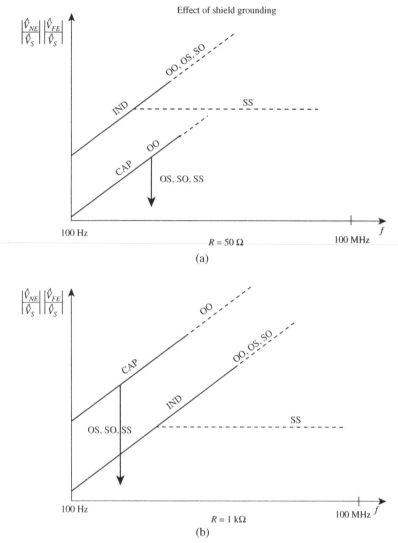

FIGURE 5.16. Explanation of the effect of shield grounding in the experimental results of Fig. 5.14 in terms of inductive and capacitive coupling for (a) $R = 50\ \Omega$ and $R = 1\ k\Omega$.

suppose that we strip back a very short section of 8 cm at each end. Our intuition tells us that the 8-cm pigtail lengths will be inconsequential since they represent 0.044% of the total shield length! We will show data that proves that this assumption is false. Here we get caught in a "sounds right" in a misleading "word game." This seeming contradiction can easily be explained by observing that "over the very long shield, the crosstalk is reduced dramatically, so that even though the pigtail lengths are short, they contribute a relatively larger contribution to the total crosstalk."

Effect of shield pigtails

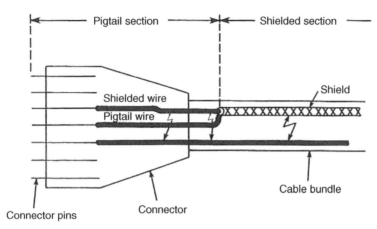

FIGURE 5.17. "Pigtails" used to terminate shields in a cable connector.

Figure 5.18 shows how to visualize and compute the effect of the pigtails on the total crosstalk. The total crosstalk is the superposition (sum) of the contributions along the two pigtails and along the long shielded section as

$$\hat{V}_{NE} = \hat{V}_{NE}^{\text{left pigtail}} + \hat{V}_{NE}^{\text{shield}} + \hat{V}_{NE}^{\text{right pigtail}} \tag{5.39}$$

5.5.1   Experimental Results

To demonstrate the effects of pigtails on crosstalk we use the previous experimental setup but with the various pigtail lengths shown in Fig. 5.19, again using two sets of terminations; the source resistance is zero, $R_S = 0$, and all the other terminations are identical resistors, R. We investigate two terminations: $R = 50\,\Omega$ (inductive coupling is dominant) and $R = 1\,\text{k}\Omega$ (capacitive coupling is dominant). Both ends of the shield via the pigtail will be grounded, SS. Three lengths of pigtails are examined: 8, 3, and 0.5 cm.

In Fig 5.20(a), for $R = 50\,\Omega$, we see that the 8- and the 5-cm pigtails begin to degrade the shielding effectiveness of the shield at around 100 kHz, so it is clear that pigtails can have a devastating effect on crosstalk reduction. What we are seeing in Fig. 5.20(a) is that the crosstalk above 100 kHz is simply the crosstalk due to the two pigtails; the crosstalk due to the long shield is dominated by them even though the crosstalk due to them is quite small. Figure 5.20(b) shows this for $R = 1\,\text{k}\Omega$. The same occurs here except that the crosstalk is due to the capacitive crosstalk of the pigtails.

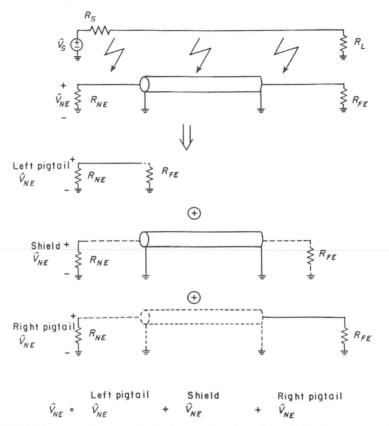

FIGURE 5.18. Approximate method of computing the effect of pigtails on crosstalk by superimposing the inductive and capacitive coupling over each section.

Figure 5.21 shows the separation of the components of their three components for $R = 50\,\Omega$. This clearly shows that above 100 kHz the pigtail contributions (inductive) dominate the contribution along the shield (inductive).

5.6   EFFECT OF MULTIPLE SHIELDS

We will briefly investigate the use of multiple shields: one on the generator line and one on the receptor line, as illustrated in Fig. 5.22.

5.6.1   Experimental Results

The cross-sectional dimensions are shown in Fig. 5.23. The result is a product of the shield factors of each shield, as can be verified as before:

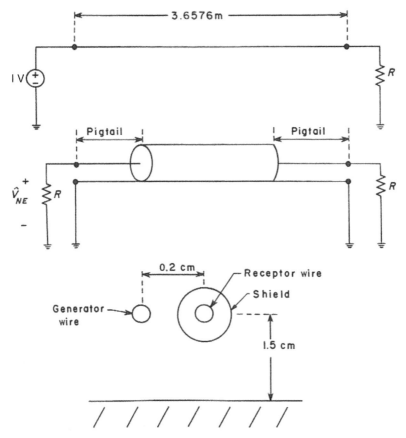

FIGURE 5.19. Experiment to illustrate the effect of pigtail lengths on crosstalk.

$$\frac{\hat{V}_{NE,FE}^{IND}}{\hat{V}_S} = \underbrace{\frac{\hat{V}_{NE,FE}^{IND}}{\hat{V}_S}}_{\substack{\text{with both} \\ \text{shields} \\ \text{removed}}} \frac{R_{SHG}}{R_{SHG} + j\omega L_{SHG}} \frac{R_{SHR}}{R_{SHR} + j\omega L_{SHR}} \qquad (5.40)$$

On the following graphs the notation "SEPARATION: WIDE" means a separation of 1.5 cm, as in Fig. 5.23. The shield break frequencies for each shield is $f_{SHG} = f_{SHR} \cong 6$ kHz. The pigtail lengths are 8 cm, and the total line length is 3.6576 m.

Figure 5.24 shows the results for three shield situations: unshielded to unshielded (no shield on either line), unshielded to shielded (one shield

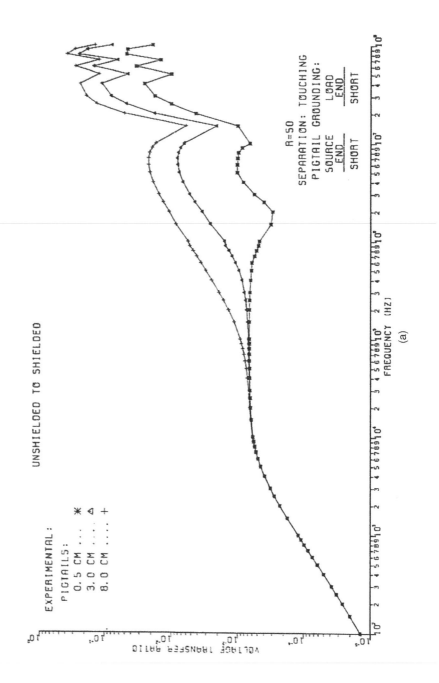

(a)

190

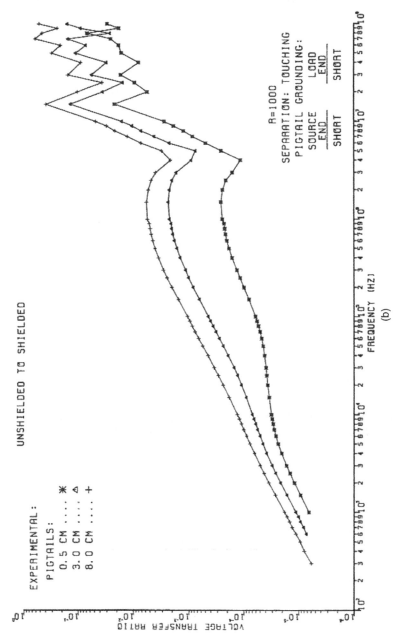

FIGURE 5.20. Experimental results for the configuration of Fig 5.19 for pigtail lengths of 0.5, 3, and 8 cm for (a) $R = 50\,\Omega$ and (b) $R = 1\,\mathrm{k}\Omega$.

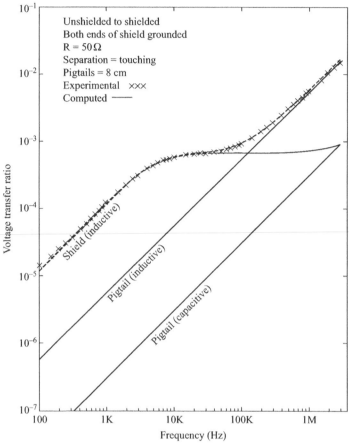

FIGURE 5.21. The near-end crosstalk for the configuration of Fig. 5.19 for 8-cm pigtail lengths and $R = 50\,\Omega$ in terms of contributions according to Fig. 5.18.

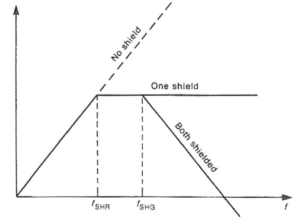

FIGURE 5.22. Frequency response of crosstalk for no shields, one shield (on the generator wire or on the receptor wire), and shields on both wires.

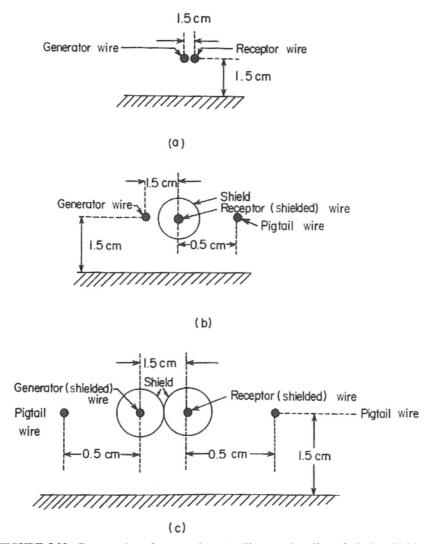

FIGURE 5.23. Cross section of an experiment to illustrate the effect of placing shields on neither wire [(a) unshielded to unshielded], one wire [(b) unshielded to shielded], or both wires [(c) shielded to shielded].

on either line), and shielded to shielded (shields on both lines). Note the effect of the 8 cm pigtails about 80 kHz and the break frequency of approximately 6 kHz.

Figure 5.25(a) and (b) show a comparison of the experimental results versus the prediction of the MTL transmission-line model. These show the excellent predictions of the MTL transmission-line model.

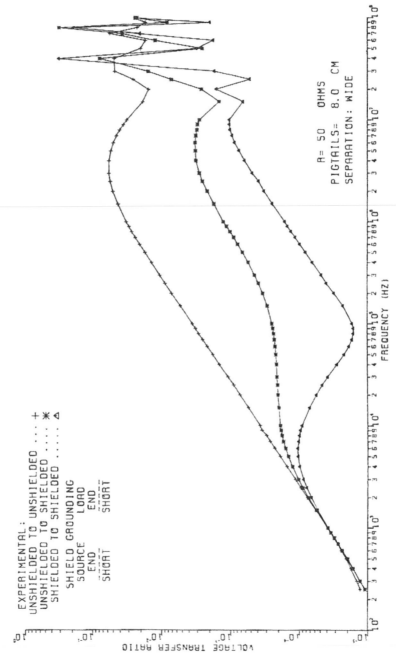

FIGURE 5.24. Experimental results for $R = 50\,\Omega$ and 8-cm pigtails for the configurations of Fig. 5.23.

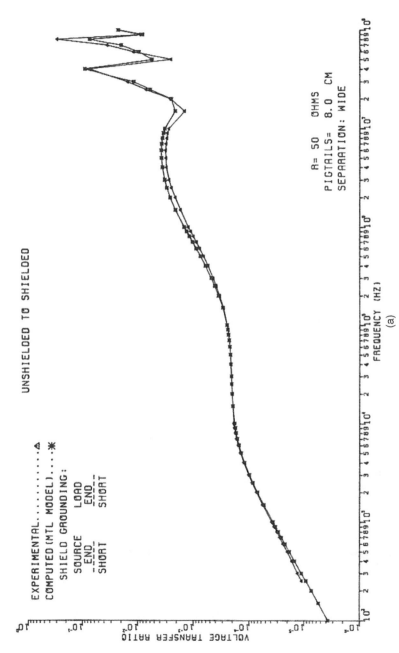

FIGURE 5.25. Predictions of the transmission-line model versus experimental results for the configuration of Fig. 5.23 for (a) a shield on the receptor wire and (b) shields on both the generator and receptor wires. (*Continued*)

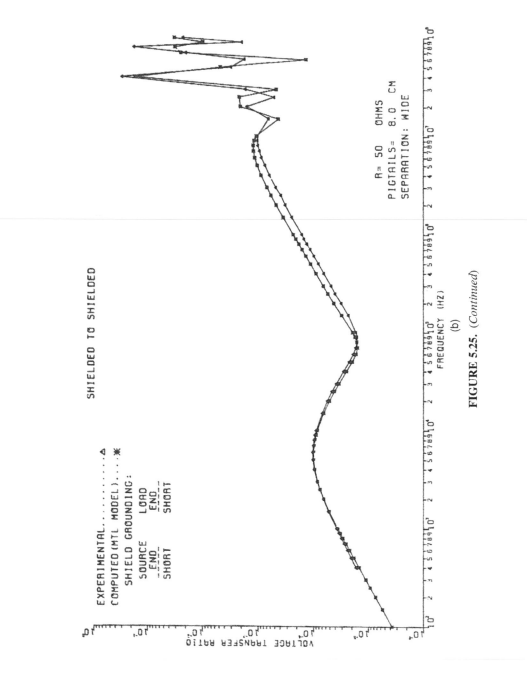

FIGURE 5.25. (*Continued*)

5.7   EFFECT OF TWISTED PAIRS OF WIRES IN REDUCING CROSSTALK

A twisted pair of wires [a twisted-wire pair (TWP)] is in reality a bifilar helix, as shown in Fig. 5.26. The effect of the twist is to *cancel out the inductive coupling due to the magnetic field of the generator circuit current, I_G (or any other current parallel to the TWP)*! Notice that the magnetic field due to I_G induces in *adjacent loops*, by Faraday's law, voltage sources in the loops which have *the same polarity in adjacent loops*. But those induced voltage sources in adjacent loops *cancel because of the twist.* Hence

> the twisted pair of wires has the important consequence of eliminating the inductive coupling.

We refer to the length of each loop as a half-twist and denote it as $\mathscr{L}_{HT}$. Hence the total length of the entire twisted wire line is $\mathscr{L} = 2N\mathscr{L}_{HT}$, where the line contains a total of $2N$ loops (half-twists). This shows that *if the TWP has an even number of half-twists, the inductive coupling is eliminated completely because the voltage sources in adjacent loops cancel,* but *if the TWP has an odd number of half-twists, the inductive coupling is that of only one half-twist of length $\mathscr{L}_{HT}$.* Hence for an odd number of half-twists, the inductive coupling is due to a very short line of length $\mathscr{L}_{HT}$, thereby reducing it dramatically but not totally eliminating it as an even number of half-twists would do.

Figure 5.27 shows an equivalent circuit of a TWP. We have shown the TWP as being over a ground plane. Although the TWP in reality rotates "smoothly," to simplify the calculations we assume that the loops (1) rotate abruptly, and

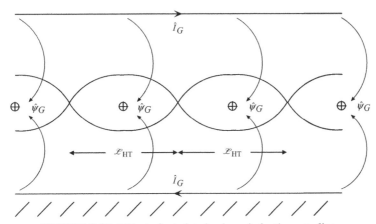

FIGURE 5.26. Effect of a twisted pair on inductive coupling.

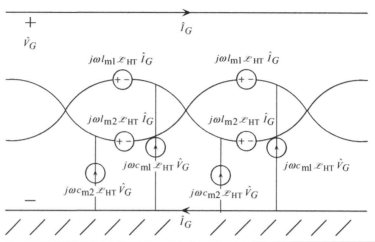

FIGURE 5.27. Equivalent circuit of a twisted pair of wires.

(2) are perpendicular to the ground plane. Figure 5.28 shows the final approximation, where the actual smooth loops are replaced by rectangular loops. A large number of experimental results and calculations have shown this to give accurate results.

The per-unit-length mutual inductances between the generator circuit and each of the wires of the TWP loops are l_{m1} and l_{m2}, and the per-unit-length mutual capacitances between the generator circuit and each of the wires of the TWP loops are c_{m1} and c_{m2}. Each "loop" is between the conductor and the ground plane. We represent these in the usual way as voltage sources and current sources. To further simplify the calculations, we represent each loop as

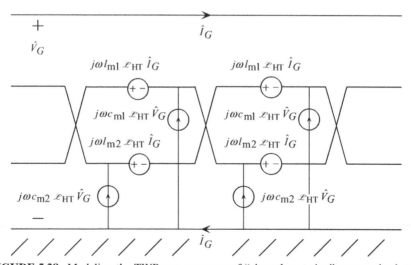

FIGURE 5.28. Modeling the TWP as a sequence of "abruptly rotating" rectangular loops.

being "rectangular" with the top and bottom sides parallel to the ground plane, as shown in Fig. 5.28. From our previous calculations we can obtain these per-unit-length parameters as each segment forming a loop between the wire and the ground plane:

$$l_{m1} = \frac{\mu_0}{4\pi} \ln\left(1 + 4\frac{h_G h_1}{s^2}\right) \tag{5.41a}$$

$$l_{m2} = \frac{\mu_0}{4\pi} \ln\left(1 + 4\frac{h_G h_2}{s^2}\right) \tag{5.41b}$$

where h_1 and h_2 are the heights of each wire above the ground plane, h_G is the height of the generator wire above the ground plane, and s is the separation between the generator wire and each of the TWP wires. Note that the wires are separated by their insulations so that $h_1 - h_2 = 2(r_w + t_w)$ and r_w and t_w are the (identical) wire radii and insulation thicknesses, respectively, The per-unit-length capacitances c_{m1} and c_{m2} are computed in the usual way (but are a bit more complicated) from

$$\begin{bmatrix} c_G + c_{m1} + c_{m2} & -c_{m1} & -c_{m2} \\ -c_{m1} & c_1 + c_{m1} + c_{m12} & -c_{m12} \\ -c_{m2} & -c_{m12} & c_2 + c_{m2} + c_{m12} \end{bmatrix}$$

$$= \mu_0 \varepsilon_0 \begin{bmatrix} l_G & l_{m1} & l_{m2} \\ l_{m1} & l_1 & l_{m12} \\ l_{m2} & l_{m12} & l_2 \end{bmatrix}^{-1} \tag{5.42}$$

The near- and far-end crosstalk is determined not only by the model of the line but also by the termination configuration. Figure 5.29 shows the untwisted line and the twisted line, but with *unbalanced terminations*. One wire of the pair is connected or "grounded" to the ground plane at only one end to avoid any "ground loops." The particular wire that is "grounded" at the left end causes that particular c_{m1} or c_{m2} to be "shorted out," leaving the capacitances as being due to the other, both of which are approximately the same: $c_{m1} \cong c_{m2} \cong c_m$. From Fig. 5.29 we obtain the near- and far-end crosstalk equations as

$$\hat{V}_{NE} = j\omega \left[\frac{R_{NE}}{(R_{NE} + R_{FE})(R_S + R_L)}(l_{m1} - l_{m2})\mathscr{L} \right.$$
$$\left. + \frac{R_{NE}R_{FE}R_L}{(R_{NE} + R_{FE})(R_S + R_L)}c_m\mathscr{L} \right] \hat{V}_S \tag{5.43a}$$

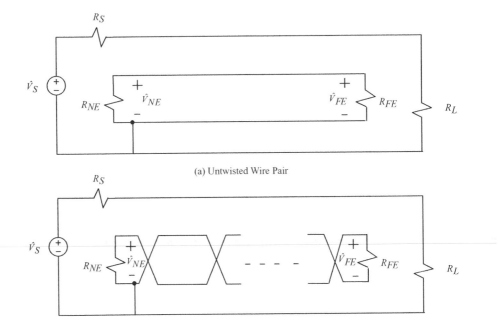

(a) Untwisted Wire Pair

(b) Twisted Wire Pair

FIGURE 5.29. Untwisted vs. twisted pair with unbalanced terminations.

$$
\hat{V}_{FE} = j\omega \left[-\frac{R_{FE}}{(R_{NE} + R_{FE})(R_S + R_L)} (l_{m1} - l_{m2})\mathscr{L} \right. \\
\left. + \frac{R_{NE}R_{FE}R_L}{(R_{NE} + R_{FE})(R_S + R_L)} c_m\mathscr{L} \right] \hat{V}_S
$$

(5.43b)

for the untwisted pair in Fig. 5.29(a) and

$$
\hat{V}_{NE} = j\omega \left[\frac{R_{NE}}{(R_{NE} + R_{FE})(R_S + R_L)} (l_{m1} - l_{m2})\mathscr{L}_{HT} \right. \\
\left. + \frac{R_{NE}R_{FE}R_L}{(R_{NE} + R_{FE})(R_S + R_L)} c_m\mathscr{L} \right] \hat{V}_S
$$

(5.44a)

$$
\hat{V}_{FE} = j\omega \left[-\frac{R_{FE}}{(R_{NE} + R_{FE})(R_S + R_L)} (l_{m1} - l_{m2})\mathscr{L}_{HT} \right. \\
\left. + \frac{R_{NE}R_{FE}R_L}{(R_{NE} + R_{FE})(R_S + R_L)} c_m\mathscr{L} \right] \hat{V}_S
$$

(5.44b)

for the twisted pair in Fig. 5.29(b). The only difference between the crosstalk results for the untwisted case and the twisted case is that inductive coupling is for the entire line for the untwisted case but for only one half-twist (for an odd number of half-twists) and zero for an even number of half-twists for the twisted pair case. For either case the per-unit-length capacitance depends on which wire of the wire pair is attached to the ground or "grounded" and is approximately the same as either $c_{m1}\mathscr{L}$ or $c_{m2}\mathscr{L}$, which we denoted as $c_m\mathscr{L}$. So unlike the inductive coupling the capacitive crosstalk is not affected by the twist. Figure 5.30 gives a summary of these results.

The final terminal configuration is that of *balanced terminations. Balanced termination means that each wire sees the same impedance from that wire to ground.* Figure 5.31 shows a balanced termination for a twisted wire pair. The balancing of the termination provides that the current sources representing the capacitive crosstalk are subtracted or, like the voltage sources in the inductive crosstalk, approximately canceled.

Hence twisted-wire pairs eliminate the inductive crosstalk, whereas balancing of the terminations eliminates the capacitive crosstalk.

The resulting crosstalk is (for an odd number of half-twists)

$$\hat{V}_{NE} = j\omega \left[\frac{R_{NE}}{(R_{NE} + R_{FE})(R_S + R_L)} (l_{m1} - l_{m2})\mathscr{L}_{HT} \right.$$
$$\left. + \frac{R_{NE}R_{FE}R_L}{(R_{NE} + R_{FE})(R_S + R_L)} (c_{m1} - c_{m2})\mathscr{L}_{HT} \right] \hat{V}_S \qquad (5.45a)$$

$$\hat{V}_{FE} = j\omega \left[-\frac{R_{FE}}{(R_{NE} + R_{FE})(R_S + R_L)} (l_{m1} - l_{m2})\mathscr{L}_{HT} \right.$$
$$\left. + \frac{R_{NE}R_{FE}R_L}{(R_{NE} + R_{FE})(R_S + R_L)} (c_{m1} - c_{m2})\mathscr{L}_{HT} \right] \hat{V}_S \qquad (5.45b)$$

Figure 5.32 shows a common implementation of an unbalanced-to-balanced termination using a center-tapped transformer referred to as a *balun* ("balanced-to-unbalanced") for making a "single-ended device" to a balanced line connection.

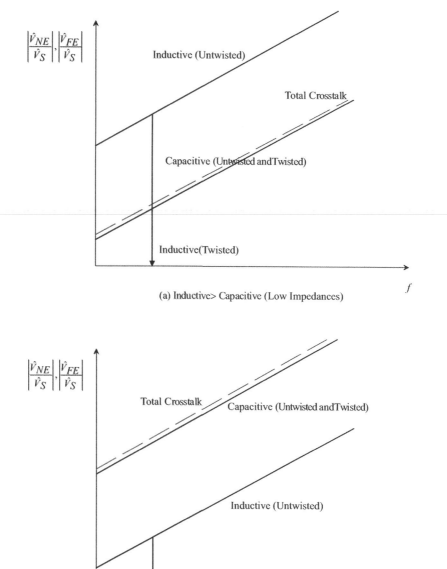

FIGURE 5.30. Summary of the crosstalk results for untwisted and twisted pairs of wires and unbalanced loads.

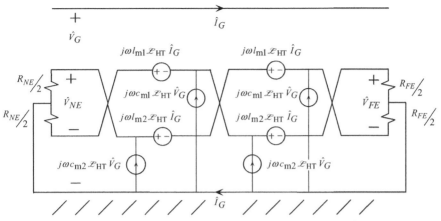

FIGURE 5.31. Balanced terminations and twisted-wire pairs.

5.7.1   Experimental Results

Figure 5.33 shows the dimensions for the experiment to investigate the effect of a twisted pair on crosstalk for unbalanced terminations. The total line length is 4.705 m. The wires are No. 20 gauge solid, separated by 2 cm, and suspended above the ground plane at a height of 2 cm. The number of half-twists is $HT = 225$, so the length of a half-twist is $\mathcal{L}_{HT} = 2.09$ cm. The cross-sectional dimensions are shown in Fig. 5.34. Figure 5.35(a) to (c) compare the straight-wire pair (SWP) (untwisted pair) and the twisted wire pair (TWP) for $R = 1\,k\Omega$, $R = 50\,\Omega$, and $R = 1\,\Omega$, respectively.

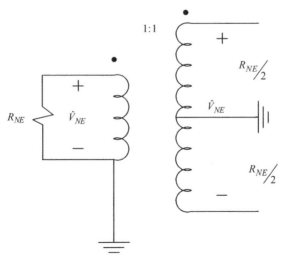

FIGURE 5.32. Practical method for balancing a termination.

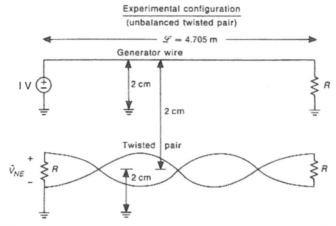

FIGURE 5.33. Experiment to illustrate the effect of a twisted pair on crosstalk for unbalanced terminations.

Figure 5.35(a) shows that for high-impedance terminations, $R = 1\ \text{k}\Omega$, a twisted pair has NO effect. Clearly, the capacitive coupling dominates the inductive coupling. The MTL transmission-line model (noted on the graph as "chain parameter model") gives very accurate predictions. For low-impedance

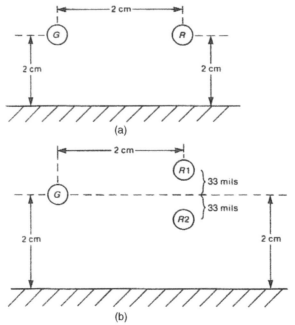

FIGURE 5.34. Cross-sectional dimensions for the experiment of Fig. 5.33: (a) single-receptor wire; (b) SWP and TWP.

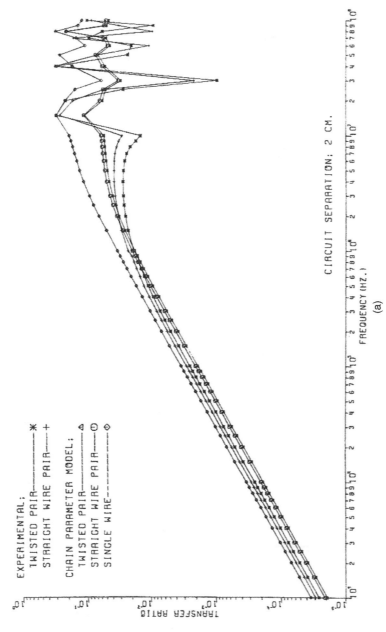

FIGURE 5.35. Experimental results for the experiment of Fig. 5.33, comparing the predictions of the transmission-line model (chain parameter model) to measured data for a single-receptor wire, a straight-wire pair, and a twisted-wire receptor pair, for (a) $R = 1 \text{k}\Omega$ (b) $R = 50 \Omega$, and (c) $R = 1 \Omega$. (*Continued*)

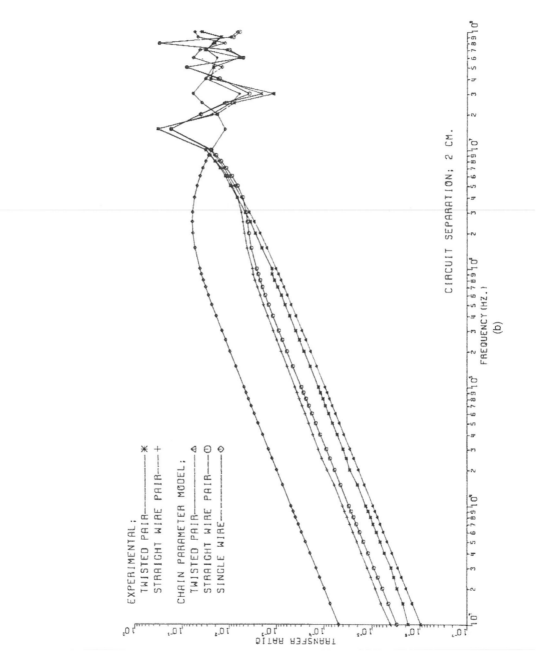

EXPERIMENTAL;
TWISTED PAIR——— ✳
STRAIGHT WIRE PAIR— +

CHAIN PARAMETER MODEL;
TWISTED PAIR——— △
STRAIGHT WIRE PAIR— ⊕
SINGLE WIRE——————— ◇

TRANSFER RATIO

FREQUENCY (HZ.)

(b)

CIRCUIT SEPARATION; 2 CM.

206

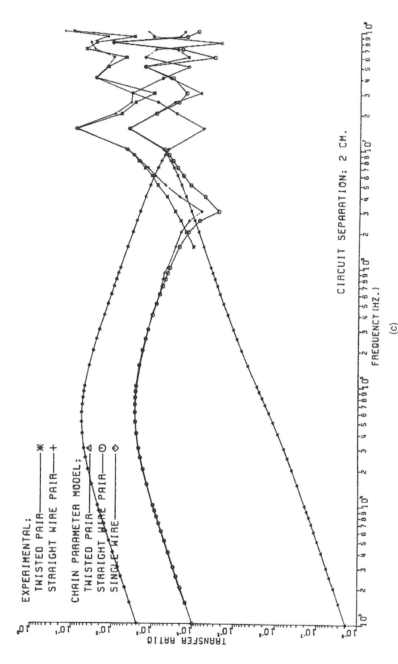

FIGURE 5.35. (*Continued*)

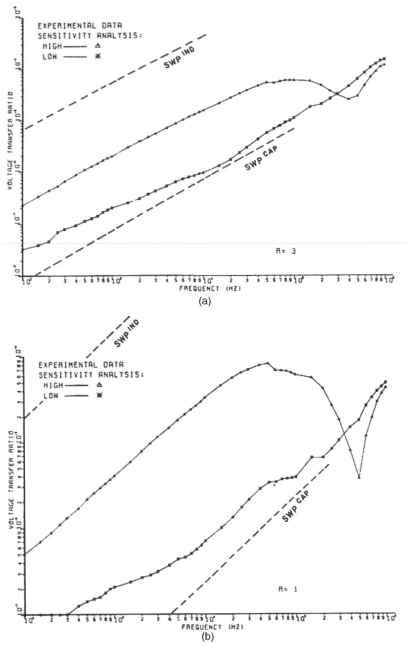

FIGURE 5.36. Experimental results for the experiment of Fig. 5.33 obtained by varying the number of twists, showing that for low-impedance loads the inductive coupling can be dominant for an odd number of half-twists and eleminated for an even number: (a) $R = 3\,\Omega$; (b) $R = 1\,\Omega$.

terminations, $R = 50\,\Omega$, shown in Fig. 5.35(b), the twisted wires have more of an effect. For VERY low-impedance terminations, $R = 1\,\Omega$, shown in Fig. 5.35 (c), the twisted wires have a more dramatic effect.

Finally, there is the question of sensitivity with TWPs. For example, a twisted-wire pair having an odd number of half-twists will have crosstalk (although very minimal) whereas a twisted-wire pair having an even number of half-twists will, theoretically, have zero crosstalk. This suggests an extreme sensitivity with a TWP. To investigate this, we constructed a generator wire to TWP over a ground plane and used unbalanced terminations, as shown in Fig. 5.33. One end was grounded but the other end was not, so we devised a miniature rotary device at the ungrounded end of the TWP. An assistant at the far end was instructed to rotate the ungrounded end and then was told to stop by the person viewing the near-end receiver at the end of the TWP that was grounded. This was done at every frequency for the "low" and "high" readings and between 100 Hz and 1 MHz and plotted in Fig 5.36. We surmised that "low" was where there was an even number of half-twists, and "high" was where there was an odd number of half-twists: a net of one half-twist. Hence going from zero recorded crosstalk to that of "something" (namely, a half-twist) is an infinite change, theoretically. Note the extreme sensitivity to twist for low-impedance loads of $3\,\Omega$ in Fig. 5.36(a) and $1\,\Omega$ in Fig. 5.36(b). This extreme sensitivity of a TWP to twist does not seem to be recognized in the EMC community, although it clearly exists, as shown by the experimental results in Fig. 5.36.

5.8   THE SHIELDED TWISTED-PAIR WIRE: THE BEST OF BOTH WORLDS

If a shield that is grounded at only one end to avoid "ground loops" is placed around a TWP that has unbalanced terminations, the capacitive coupling is eliminated by the single-end-grounded shield, and the inductive coupling is eliminated by the unbalanced TWP, which eliminates the total crosstalk. This is the simplest and most effective method for reducing crosstalk.

6

THE EXACT CROSSTALK PREDICTION MODEL

In this chapter we determine a method for the *exact* numerical solution of the transmission-line equations for a three-conductor *lossless line* and the resulting crosstalk. Solution of the transmission-line equations for any *lossless multiconductor line* is virtually *trivial*, as we will see in this chapter. A SPICE (PSPICE) subcircuit model will be developed for any *lossless* MTL having any number of parallel conductors. Terminations (linear or nonlinear) can then be attached to that PSPICE subcircuit model, and the resulting crosstalk can be determined easily and accurately. Unlike the inductive–capacitive approximate coupling model in Chapter 5, it gives only a numerical solution and does not supply the intuition as to what causes the crosstalk that is provided with the inductive–capacitice coupling model of Chapter 5.

For *three-conductor, lossless transmission lines*, the first-order transmission-line equations that describe crosstalk between the two transmission lines are *coupled* and of the form

$$\frac{\partial \mathbf{V}(z, t)}{\partial z} = -\mathbf{L}\frac{\partial \mathbf{I}(z, t)}{\partial t} \tag{6.1a}$$

Transmission Lines in Digital Systems for EMC Practitioners, First Edition.
Clayton R. Paul.
© 2012 John Wiley & Sons, Inc. Published 2012 by John Wiley & Sons, Inc.

$$\boxed{\frac{\partial \mathbf{I}(z,t)}{\partial z} = -\mathbf{C}\frac{\partial \mathbf{V}(z,t)}{\partial t}}$$

(6.1b)

where the 2×1 vectors of line voltages and line currents have entries that are functions of position along the line, z, and time, t:

$$\mathbf{V}(z,t) = \begin{bmatrix} V_G(z,t) \\ V_R(z,t) \end{bmatrix}$$

(6.2a)

$$\mathbf{I}(z,t) = \begin{bmatrix} I_G(z,t) \\ I_R(z,t) \end{bmatrix}$$

(6.2b)

The entries in the 2×2 per-unit-length inductance and capacitance matrices, **L** and **C**, respectively, are *independent* of position along the line, z, and time, t, and contain all the cross-sectional dimensions and properties of the line. These are of the form

$$\mathbf{L} = \begin{bmatrix} l_G & l_m \\ l_m & l_R \end{bmatrix}$$

(6.3a)

and

$$\mathbf{C} = \begin{bmatrix} c_G + c_m & -c_m \\ -c_m & c_R + c_m \end{bmatrix}$$

(6.3b)

6.1   DECOUPLING THE TRANSMISSION-LINE EQUATIONS WITH MODE TRANSFORMATIONS

The key to solving the *coupled* MTL equations in (6.1) is to *simultaneously decouple them* with *similarity transformations*. Define a *change of variables* to convert the *actual* line voltages and *actual* line currents to the *mode* voltages and *mode* currents as

$$\boxed{\mathbf{V}(z,t) = \mathbf{T}_V \, \mathbf{V}_{\mathrm{mode}}(z,t)}$$

(6.4a)

$$\boxed{\mathbf{I}(z, t) = \mathbf{T}_I \, \mathbf{I}_{\text{mode}}(z, t)} \tag{6.4b}$$

Observe that the 2×2 mode transformations $\mathbf{T}_V$ and $\mathbf{T}_I$ are *independent* of position along the line, z, and time, t. These vectors of *mode voltages* and *mode currents* are of the form

$$\mathbf{V}_{\text{mode}}(z, t) = \begin{bmatrix} V_{mG}(z, t) \\ V_{mR}(z, t) \end{bmatrix} \tag{6.5a}$$

$$\mathbf{I}_{\text{mode}}(z, t) = \begin{bmatrix} I_{mG}(z, t) \\ I_{mR}(z, t) \end{bmatrix} \tag{6.5b}$$

and the 2×2 mode tranformations have entries

$$\mathbf{T}_V = \begin{bmatrix} T_{VGG} & T_{VGR} \\ T_{VRG} & T_{VRR} \end{bmatrix} \tag{6.6a}$$

$$\mathbf{T}_I = \begin{bmatrix} T_{IGG} & T_{IGR} \\ T_{IRG} & T_{IRR} \end{bmatrix} \tag{6.6b}$$

Now substitute the mode transformations in (6.4) into the MTL equations to give

$$\boxed{\frac{\partial \mathbf{V}_{\text{mode}}}{\partial z} = - \underbrace{\mathbf{T}_V^{-1} \mathbf{L} \mathbf{T}_I}_{\mathbf{L}_m} \frac{\partial \mathbf{I}_{\text{mode}}}{\partial t}} \tag{6.7a}$$

$$\boxed{\frac{\partial \mathbf{I}_{\text{mode}}}{\partial z} = - \underbrace{\mathbf{T}_I^{-1} \mathbf{C} \mathbf{T}_V}_{\mathbf{C}_m} \frac{\partial \mathbf{V}_{\text{mode}}}{\partial t}} \tag{6.7b}$$

If we can choose the 2×2 mode transformations $\mathbf{T}_V$ and $\mathbf{T}_I$ such that $\mathbf{L}$ and $\mathbf{C}$ are *simultaneously* diagonalized as

$$\mathbf{L}_m = \mathbf{T}_V^{-1} \mathbf{L} \mathbf{T}_I$$

$$= \begin{bmatrix} l_{mG} & 0 \\ 0 & l_{mR} \end{bmatrix} \tag{6.8a}$$

and

$$\mathbf{C}_m = \mathbf{T}_I^{-1} \mathbf{C} \mathbf{T}_V$$

$$= \begin{bmatrix} c_{mG} & 0 \\ 0 & c_{mR} \end{bmatrix} \tag{6.8b}$$

then the MTL equations are decoupled into two sets of *uncoupled* two-conductor mode lines as

$$\frac{\partial V_{mG}(z, t)}{\partial z} = -l_{mG} \frac{\partial I_{mG}(z, t)}{\partial t}$$

$$\frac{\partial I_{mG}(z, t)}{\partial z} = -c_{mG} \frac{\partial V_{mG}(z, t)}{\partial t} \tag{6.9a}$$

and

$$\frac{\partial V_{mR}(z, t)}{\partial z} = -l_{mR} \frac{\partial I_{mR}(z, t)}{\partial t}$$

$$\frac{\partial I_{mR}(z, t)}{\partial z} = -c_{mR} \frac{\partial V_{mR}(z, t)}{\partial t} \tag{6.9b}$$

giving two *uncoupled* and independent sets of two-conductor mode transmission-line equations having characteristic impedances of *each mode* as

$$Z_{CmG} = \sqrt{\frac{l_{mG}}{c_{mG}}}$$

$$Z_{CmR} = \sqrt{\frac{l_{mR}}{c_{mR}}} \tag{6.10}$$

and velocities of propagation of *each mode* as

$$v_{mG} = \frac{1}{\sqrt{l_{mG}\, c_{mG}}}$$
$$v_{mR} = \frac{1}{\sqrt{l_{mR}\, c_{mR}}} \tag{6.11}$$

Note that *each of these decoupled mode transmission lines* can be represented by a two-conductor line and can be simulated using the built-in PSPICE two-conductor line model! From now on we omit showing the dependence of the line and mode voltages and currents on position along the line, z, and time, t, with that implicit understanding.

6.2 THE SPICE SUBCIRCUIT MODEL

But how do we get back from the mode voltages and currents to the actual line voltages and currents? The answer is to build a PSPICE model using *controlled sources* that simulates these mode transformations. The mode transformations in (6.4) can be written as

$$\mathbf{V} = \mathbf{T}_V \mathbf{V}_{\text{mode}} \tag{6.12a}$$

$$\mathbf{I}_{\text{mode}} = \mathbf{T}_I^{-1} \mathbf{I} \tag{6.12b}$$

or

$$V_G = T_{VGG} V_{mG} + T_{VGR} V_{mR}$$
$$V_R = T_{VRG} V_{mG} + T_{VRR} V_{mR} \tag{6.13a}$$

and

$$I_{mG} = T_{IGG}^{-1} I_G + T_{IGR}^{-1} I_R$$
$$I_{mR} = T_{IRG}^{-1} I_G + T_{IRR}^{-1} I_R \tag{6.13b}$$

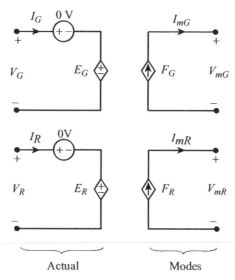

Actual   Modes

FIGURE 6.1. Equivalent circuit giving the transformation between the actual line voltages and currents and the mode voltages and currents.

Note that $T_{IGG}^{-1} \neq 1/T_{IGG}$; it just symbolizes the entry in $\mathbf{T}_I^{-1}$. The equivalent circuit that gives this transformation is shown in Fig. 6.1, where

$$
\begin{aligned}
E_G &= T_{VGG}V_{mG} + T_{VGR}V_{mR} \\
E_R &= T_{VRG}V_{mG} + T_{VRR}V_{mR}
\end{aligned}
\tag{6.14a}
$$

and

$$
\begin{aligned}
F_G &= T_{IGG}^{-1}I_G + T_{IGR}^{-1}I_R \\
F_R &= T_{IRG}^{-1}I_G + T_{IRR}^{-1}I_R
\end{aligned}
\tag{6.14b}
$$

The complete PSPICE subcircuit model is shown in Fig. 6.2.

But how do we determine the mode transformations $\mathbf{T}_V$ and $\mathbf{T}_I$ that will *simultaneously diagonalize* $\mathbf{L}$ and $\mathbf{C}$? For 2×2 $\mathbf{L}$ and $\mathbf{C}$, this is simple. There are some important, basic facts about similarity transformations of matrices.

1. $\mathbf{L}$ and $\mathbf{C}$ are both real and symmetric (i.e., their entries are real numbers and $L_{12} = L_{21}$ and $C_{12} = C_{21}$).
2. Any real symmetric matrix

$$
\mathbf{M} = \begin{bmatrix} M_{11} & M_{12} \\ M_{12} & M_{22} \end{bmatrix}
$$

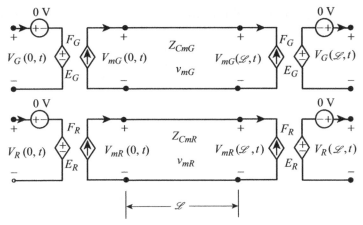

FIGURE 6.2. Complete PSPICE subcircuit model.

can be diagonalized with a real *orthogonal transformation*

$$T = \begin{bmatrix} T_{11} & T_{12} \\ T_{21} & T_{22} \end{bmatrix}$$

such that

$$T^{-1}MT = T^{t}MT$$
$$= \begin{bmatrix} D_1 & 0 \\ 0 & D_2 \end{bmatrix} \tag{6.15}$$

where the inverse of **T** is its *transpose*:

$$T^{-1} = T^{t} \tag{6.16}$$

and T^{t} denotes its *transpose* (rotation about its main diagonal), $\lfloor T^{t} \rfloor_{ij} = [T]_{ji}$. In fact, this is simple for a 2×2 real symmetric ($M_{12} = M_{21}$) matrix:

$$M = \begin{bmatrix} M_{11} & M_{12} \\ M_{12} & M_{22} \end{bmatrix}$$

and the real orthogonal transformation matrix $\mathbf{T}$ is simple to obtain:

$$\mathbf{T} = \begin{bmatrix} \cos\theta & -\sin\theta \\ \sin\theta & \cos\theta \end{bmatrix} \tag{6.17a}$$

where angle θ is obtained from

$$\tan(2\theta) = \frac{2M_{12}}{M_{11} - M_{22}} \tag{6.17b}$$

EXAMPLE

$$\mathbf{M} = \begin{bmatrix} 5 & 2 \\ 2 & 3 \end{bmatrix}$$

$$\tan(2\theta) = \frac{4}{5-3}$$

so that $\theta = 31.72°$. So

$$\mathbf{T} = \begin{bmatrix} 0.85 & -0.53 \\ 0.53 & 0.85 \end{bmatrix}$$

Note that the *columns* of $\mathbf{T}$ can be thought of as vectors in two-dimensional space and, by this procedure, are automatically normalized to a length (in space) of unity (i.e., $\sqrt{\cos^2\theta + \sin^2\theta} = 1$). They are also perpendicular to each other in space. Hence it is said to be an *orthogonal transformation.* You can check that this $\mathbf{T}$ has the properties

$$\mathbf{T}^{-1} = \mathbf{T}^t$$

$$\mathbf{T}^{-1}\mathbf{MT} = \mathbf{T}^t\mathbf{MT} = \begin{bmatrix} 6.24 & 0 \\ 0 & 1.76 \end{bmatrix}$$

Case 1: Homogeneous Media, $\mathbf{L} = (1/v^2)\mathbf{C}^{-1}$ First diagonalize $\mathbf{L}$ with the above method to give

$$\mathbf{T}^{-1}\mathbf{L}\mathbf{T} = \mathbf{T}^t\mathbf{L}\mathbf{T} = \begin{bmatrix} l_{mG} & 0 \\ 0 & l_{mR} \end{bmatrix}$$

But $\mathbf{L} = (1/v^2)\mathbf{C}^{-1}$ where the medium is homogeneous and

$$v = \frac{1}{\sqrt{\mu\varepsilon}}$$

Hence

$$\mathbf{T}^{-1}\mathbf{L}\mathbf{T} = \mathbf{T}^t\mathbf{L}\mathbf{T} = \begin{bmatrix} l_{mG} & 0 \\ 0 & l_{mR} \end{bmatrix}$$
$$= \frac{1}{v^2}\mathbf{T}^{-1}\mathbf{C}^{-1}\mathbf{T} = \frac{1}{v^2}\mathbf{T}^t\mathbf{C}^{-1}\mathbf{T}$$

Therefore,

$$\mathbf{T}^{-1}\mathbf{C}\mathbf{T} = \mathbf{T}^t\mathbf{C}\mathbf{T} = \begin{bmatrix} \dfrac{1}{v^2 l_{mG}} & 0 \\ 0 & \dfrac{1}{v^2 l_{mR}} \end{bmatrix}$$
$$= \begin{bmatrix} c_{mG} & 0 \\ 0 & c_{mR} \end{bmatrix}$$

so we can define

$$\boxed{\begin{aligned} \mathbf{T}_V &= \mathbf{T} \\ \mathbf{T}_I &= \mathbf{T} \end{aligned}}$$ (6.18)

The individual mode characteristic impedances are

$$\boxed{\begin{aligned} Z_{CmG} &= v l_{mG} \\ Z_{CmR} &= v l_{mR} \end{aligned}}$$ (6.19)

and for this *homogeneous medium* all mode velocities of propagation are equal:

$$v_{mG} = v_{mR} = v = \frac{1}{\sqrt{\mu\varepsilon}} \qquad (6.20)$$

Case 2: Inhomogeneous Media, $\mathbf{L} \neq (1/v^2)\mathbf{C}^{-1}$ First diagonalize $\mathbf{C}$ with an orthogonal transformation:

$$\mathbf{U}^{-1}\mathbf{C}\mathbf{U} = \mathbf{U}^t\mathbf{C}\mathbf{U} = \boldsymbol{\theta}^2 = \begin{bmatrix} \theta_1^2 & 0 \\ 0 & \theta_2^2 \end{bmatrix}$$

Since $\mathbf{C}$ is positive definite, all its eigenvalues are positive and nonzero, so that its square roots will be real and nonzero. Then form

$$\boldsymbol{\theta}\mathbf{U}^t\mathbf{L}\mathbf{U}\boldsymbol{\theta}$$

But this is real and symmetric, so find another orthogonal transformation $\mathbf{S}$ to diagonalize this:

$$\mathbf{S}^t(\boldsymbol{\theta}\mathbf{U}^t\mathbf{L}\mathbf{U}\boldsymbol{\theta})\mathbf{S} = \begin{bmatrix} \Lambda_1^2 & 0 \\ 0 & \Lambda_2^2 \end{bmatrix}$$

Then define a 2×2 matrix $\mathbf{T}$ as

$$\mathbf{T} = \mathbf{U}\boldsymbol{\theta}\mathbf{S}$$

Normalize the columns of $\mathbf{T}$ to a length of unity:

$$\mathbf{T}_{\text{norm}} = \mathbf{U}\boldsymbol{\theta}\mathbf{S}\boldsymbol{\alpha}$$
$$= \mathbf{T}\boldsymbol{\alpha}$$

where $\boldsymbol{\alpha}$ is a 2×2 diagonal matrix with main diagonal entries

$$\alpha_{11} = \frac{1}{\sqrt{T_{11}^2 + T_{21}^2}}$$

$$\alpha_{22} = \frac{1}{\sqrt{T_{12}^2 + T_{22}^2}}$$

Then define the mode transformations as

$$\boxed{\begin{aligned}\mathbf{T}_I &= \mathbf{T}_{\text{norm}} = \mathbf{U}\boldsymbol{\theta}\mathbf{S}\boldsymbol{\alpha} \\ \mathbf{T}_V &= \mathbf{U}\boldsymbol{\theta}^{-1}\mathbf{S}\boldsymbol{\alpha}^{-1}\end{aligned}}$$

(6.21)

You can check that

$$\mathbf{T}_V^{-1}\mathbf{L}\mathbf{T}_I = \boldsymbol{\alpha}\mathbf{S}^t\boldsymbol{\theta}\mathbf{U}^t\mathbf{L}\mathbf{U}\boldsymbol{\theta}\mathbf{S}\boldsymbol{\alpha}$$

$$= \begin{bmatrix} \alpha_{11}^2\Lambda_1^2 & 0 \\ 0 & \alpha_{22}^2\Lambda_2^2 \end{bmatrix}$$

and

$$\mathbf{T}_I^{-1}\mathbf{C}\mathbf{T}_V = \boldsymbol{\alpha}^{-1}\mathbf{S}^t\boldsymbol{\theta}^{-1}\mathbf{U}^t\mathbf{C}\mathbf{U}\boldsymbol{\theta}^{-1}\mathbf{S}\boldsymbol{\alpha}^{-1}$$

$$= \begin{bmatrix} \dfrac{1}{\alpha_{11}^2} & 0 \\ 0 & \dfrac{1}{\alpha_{22}^2} \end{bmatrix}$$

So

$$l_{mG} = \alpha_{11}^2\Lambda_1^2$$
$$l_{mR} = \alpha_{22}^2\Lambda_2^2$$

and

$$c_{mG} = \frac{1}{\alpha_{11}^2}$$

$$c_{mR} = \frac{1}{\alpha_{22}^2}$$

Hence the characteristic impedances of the modes are

$$\boxed{\begin{aligned} Z_{CmG} &= \sqrt{\frac{l_{mG}}{c_{mG}}} = \alpha_{11}^2\Lambda_1 \\ Z_{CmR} &= \sqrt{\frac{l_{mR}}{c_{mR}}} = \alpha_{22}^2\Lambda_2 \end{aligned}}$$

(6.22)

and the individual mode velocities of propagation are

$$
\begin{aligned}
v_{mG} &= \frac{1}{\sqrt{l_{mG}c_{mG}}} = \frac{1}{\Lambda_1} \\
v_{mR} &= \frac{1}{\sqrt{l_{mR}c_{mR}}} = \frac{1}{\Lambda_2}
\end{aligned}
\tag{6.23}
$$

The previous development allows us to develop a SPICE subcircuit model for direct insertion into a SPICE analysis in order to easily determine the crosstalk for any *lossless* multiconductor transmission-line (MTL) problem.

The SPICEMTL Computer Program (SPICEMTL.FOR, SPICEMTL. EXE) SPICEMTL.FOR is a FORTRAN program for generating the SPICE subcircuit model of the *lossless* MTL shown in Fig. 6.2. The program SPICEMTL.EXE is the compiled and executable file. SPICEMTL.EXE reads line dimensions from the file SPICEMTL.IN and then reads the **L** and **C** matrices for the line from PUL.DAT. It generates a subcircuit model of the line that is output to the file SPICEMTL.OUT. The subcircuit model shown in Fig. 6.2 is delivered to SPICEMTL.OUT and starts with the usual subcircuit line,

```
.SUBCKT MTL 101 102 201 202
```

and ends with

```
.ENDS MTL
```

The internal nodes of the subcircuit model are connected to the nodes of the external transmission-line circuit as shown in Fig. 6.3. In Fig. 6.3, the nodes of the external MTL are named S, L, NE, and FE for bookkeeping purposes. It is very important that the user connect the proper nodes in the subcircuit model to the corresponding conductor numbers in PUL.DAT computed for this structure. Here the entries in PUL.DAT were computed for $1 = G$ and $2 = R$. Hence nodes 101 and 201 are associated with the ends of the generator conductor, and nodes 102 and 202 are associated with the ends of the receptor conductor. The 0 reference node is the only node common to both the subcircuit and the external SPICE circuit. The connections are made in the SPICE program with the statement

```
XMTL S NE L FE MTL
```

This establishes the connection $S \rightarrow 101, NE \rightarrow 102, L \rightarrow 201$, and $FE \rightarrow 202$.

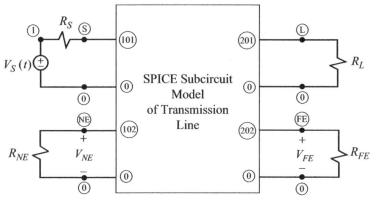

FIGURE 6.3. Connection of the nodes of the subcircuit MTL to the external nodes of the actual circuit.

EXAMPLE

Consider the example shown in Fig. 6.4. The cross-sectional configuration is a PCB having land widths of 15 mils, edge-to-edge separations of 45 mils, and board thickness of 47 mils. The board is constructed of glass epoxy having $\varepsilon_r \cong 4.7$, and the lands are of total length $10\,\text{in} = 0.254\,\text{m}$. The lands are terminated in 50-Ω resistors, and the source voltage $V_S(t)$ is a 10 MHz clock having a 1-V amplitude, a 50% duty cycle, and 6.25-ns 0 to 100% rise and fall times, which are equivalent to $\tau_{r,10-90\%} = \tau_{f,10-90\%} = 5\,\text{ns}$.

The PCB.IN data are:

3	=TOTAL NUMBER OF LANDS
30.	=NUMBER OF CONDUCTOR SUBSECTIONS
3	=NUMBER OF REFERENCE LAND
15.0	=LAND WIDTH (mils)
45.0	=EDGE-TO-EDGE LAND SEPARATION (mils)
47.0	=BOARD THICKNESS (mils)
4.7	=BOARD RELATIVE DIELECTRIC CONSTANT

The output data for L and C are in PUL.DAT:

1	1	1.38314E-06	=L(1, 1)
1	2	6.91570E-07	=L(1, 2)
2	2	1.10706E-06	=L(2, 2)

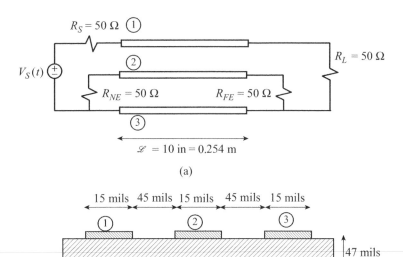

(a)

(b)

(c)

FIGURE 6.4. Example.

1	1	2.96949E-11	=C (1, 1)
1	2	-2.02619E-11	=C (1, 2)
2	2	4.05238E-11	=C (2, 2)
1	1	1.16983E-11	=CO (1, 1)
1	2	-7.30777E-12	=CO (1, 2)
2	2	1.46155E-11	=CO (2, 2)

```
NUMBER OF LANDS=  3
NUMBER OF DIVISIONS PER LAND=  30
REFERENCE LAND=  3
LAND WIDTH (mils) = 1.500E+01
```

```
EDGE-TO-EDGE SEPARATION (mils)=  4.500E+01
BOARD THICKNESS (mils)=  4.700E+01
RELATIVE DIELECTRIC CONSTANT=  4.700E+00
```

The SPICEMTL.IN data are:

```
3              =Total Number of Conductors (N+1)
0.254          =Total Line Length (m)
```

The SPICEMTL.OUT file contains the PSPICE-generated subcircuit model:

```
*SUBCIRCUIT MODEL OF A MULTICONDUCTOR TRANSMISSION
+ LINE*
*      NUMBER OF CONDUCTORS= 3
*      TOTAL LINE LENGTH (METERS)= 2.54000E-01
*      L( 1, 1)= 1.38314E-06
*      L( 1, 2)= 6.91570E-07
*      L( 2, 2)= 1.10706E-06
*      C( 1, 1)= 2.96949E-11
*      C( 1, 2)=-2.02619E-11
*      C( 2, 2)= 4.05238E-11
.SUBCKT MTL
+ 101
+ 102
+ 201
+ 202
V101     101     301
EC101     301     0     POLY( 2)
+   (501,0)
+   (502,0)
+0
+     1.000000E+00
+     -1.046409E-07
FC101     0 501     POLY( 2)
+ V101
+ V102
+0
+     1.000000E+00
+     5.000001E-01
V201     201     401
```

```
EC201     401     0     POLY( 2 )
+ (601,0)
+ (602,0)
+0
+     1.000000E+00
+     -1.046409E-07
FC201     0     601     POLY( 2 )
+ V201
+ V202
+0
+     1.000000E+00
+     5.000001E-01
V102     102     302
EC102     302     0     POLY( 2 )
+ (501,0)
+ (502,0)
+0
+     5.000001E-01
+     1.118034E+00
FC102     0     502     POLY( 2 )
+ V101
+ V102
+0
+     -1.046409E-07
+     1.118034E+00
V202     202     402
EC202 402 0 POLY( 2 )
+ (601,0)
+ (602,0)
+0
+     5.000001E-01
+     1.118034E+00
FC202     0     602     POLY( 2 )
+ V201
+ V202
+0
+     -1.046409E-07
+     1.118034E+00
T101 501 0 601 0 Z0= 2.658917E+02 TD= 1.321281E-09
T102 502 0 602 0 Z0= 1.096493E+02  TD=  1.410781E-09
.ENDS MTL
```

Adding the terminal circuitry to this subcircuit model gives the SPICE.IN file:

```
SPICE MTL MODEL; INHOMOGENEOUS MEDIUM
VS 1 0 PULSE(0 1 0 6.25N 6.25N 43.75N 100N)
*VS 1 0 AC 1 0
RS 1 S 50
RL L 0 50
RNE NE 0 50
RFE FE 0 50
XMTL S NE L FE MTL
.TRAN 0.1N 100N 0 0.1N
*.AC DEC 50 10K 1000MEG
.PROBE
*SUBCIRCUIT MODEL OF A MULTICONDUCTOR TRANSMISSION
+ LINE*
*      NUMBER OF CONDUCTORS= 3
*      TOTAL LINE LENGTH (METERS)= 2.54000E-01
*      L( 1, 1)= 1.38314E-06
*      L( 1, 2)= 6.91570E-07
*      L( 2, 2)= 1.10706E-06
*      C( 1, 1)= 2.96949E-11
*      C( 1, 2)=-2.02619E-11
*      C( 2, 2)= 4.05238E-11
.SUBCKT MTL
+ 101
+ 102
+ 201
+ 202
V101    101   301
EC101   301   0    POLY( 2)
+ (501,0)
+ (502,0)
+0
+      1.000000E+00
+      -1.046409E-07
FC101   0  501  POLY( 2)
+ V101
+ V102
+0
+      1.000000E+00
+      5.000001E-01
```

```
V201  201  401
EC201  401  0  POLY( 2)
+ (601,0)
+ (602,0)
+0
+    1.000000E+00
+   -1.046409E-07
FC201  0  601  POLY( 2)
+ V201
+ V202
+0
+    1.000000E+00
+    5.000001E-01
V102  102  302
EC102  302  0  POLY( 2)
+ (501,0)
+ (502,0)
+0
+    5.000001E-01
+    1.118034E+00
FC102  0  502  POLY( 2)
+ V101
+ V102
+0
+   -1.046409E-07
+    1.118034E+00
V202  202  402
EC202  402  0  POLY( 2)
+ (601,0)
+ (602,0)
+0
+    5.000001E-01
+    1.118034E+00
FC202  0  602  POLY( 2)
+ V201
+ V202
+0
+   -1.046409E-07
+    1.118034E+00
T101 501 0 601 0 Z0=  2.658917E+02 TD= 1.321281E-09
T102 502 0 602 0 Z0=  1.096493E+02 TD= 1.410781E-09
.ENDS MTL
.END
```

Note that there are two distinct mode velocities: $v_{m1} = \mathcal{L}/T_{D1} = 1.922 \times 10^8$ m/s and $v_{m2} = \mathcal{L}/T_{D2} = 1.8 \times 10^8$ m/s. Both of these are approximately 60% of the speed of light in free space.

Running the SPICE.IN program determines the near end and far end crosstalk voltages as shown in Fig. 6.5.

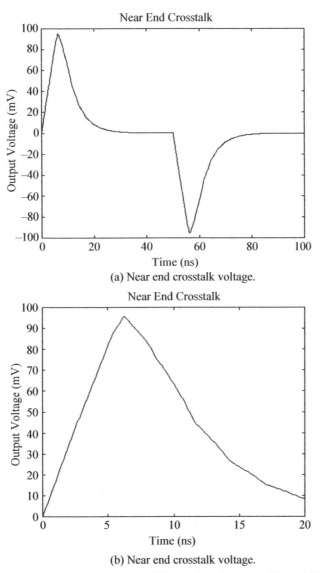

(a) Near end crosstalk voltage.

(b) Near end crosstalk voltage.

FIGURE 6.5. Near- and far-end crosstalk voltages for the problem of Fig. 6.4.

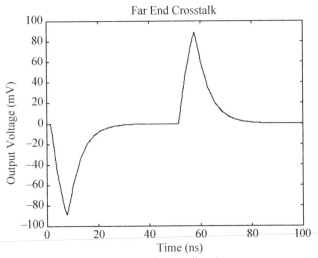

(c) Far-end crosstalk voltage.

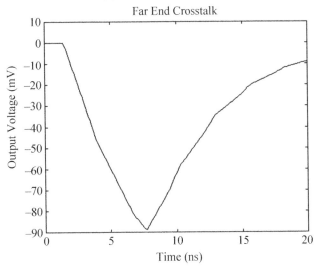

(d) Far-end crosstalk voltage.

FIGURE 6.5. (*Continued*)

These crosstalk voltages are verified, *experimentally*, in

1. C. R. Paul, *Introduction to Electromagnetic Compatibility*, 2nd ed., Wiley-Interscience, Hoboken, NJ, 2006.
2. C. R. Paul, *Analysis of Multiconductor Transmission Lines*, 2nd ed., Wiley-Interscience, Hoboken, NJ, 2008.

The near- and far-end frequency-domain transfer functions $\hat{V}_{NE}/\hat{V}_S$ and $\hat{V}_{FE}/\hat{V}_S$ can be obtained by adding

```
VS 1 0 AC 1 0
.AC DEC 50 10K 1000MEG
.PRINT AC VM(NE) VP(NE) VM(FE) VP(FE)
```

giving the magnitudes of the frequency-domain transfer functions shown in Figure 6.6.

The line (in free space with the dielectric substrate removed) is one wavelength at

$$f = \frac{v_0}{\mathscr{L}} = \frac{3 \times 10^8}{0.254} = 1.18 \, \text{GHz}$$

However, because of the inhomogeneous medium, there are two mode velocities: $v_1 = 1.92 \times 10^8$ and $v_2 = 1.8 \times 10^8$. From the plots of the transfer functions, the frequencies where the line appears to be one wavelength are around 700 MHz. The inverse of the two-mode time delays generated by the SPICE subcircuit model give these as $1/T_{D1} = 756$ MHz and $1/T_{D2} = 708$ MHz. Notice that the magnitudes of the crosstalk transfer functions increase linearly with frequency for frequencies where the line is electrically short. From the frequency-response plots in Fig. 6.6, this appears to be below 70 MHz. This linear increase of the transfer function with frequency where the line is electrically short is a general result for resistive terminatious.

6.3 LUMPED-CIRCUIT APPROXIMATE MODELS OF THE LINE

As was the case for two-conductor lines, we can construct lumped equivalent circuits of the line for frequencies of the source where the line is *electrically short*. The most common type is the lumped-Pi circuit shown in Fig. 6.7, wherein the per-unit-length self- and mutual inductances are multiplied by the line length, $\mathscr{L}$, and placed in the middle, and the per-unit-length self- and mutual capacitances are multiplied by the line length, $\mathscr{L}$, and are then split and placed on both ends. This is done to make the model symmetrical, as is the actual line.

Lumped-Circuit Approximate Models of the Line (SPICELPI.FOR, SPICELPI.EXE) The FORTRAN program SPICELPI.FOR constructs a subcircuit model for a one-section lumped-pi equivalent circuit as shown in Fig. 6.7. The executable file SPICELPI.EXE reads the line dimensions from

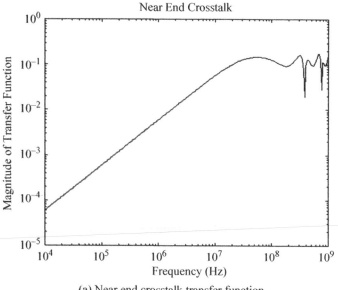

(a) Near end crosstalk transfer function.

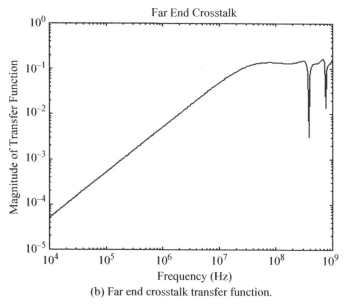

(b) Far end crosstalk transfer function.

FIGURE 6.6. Magnitudes of the frequency-domain transfer functions for the problem shown in Fig. 6.4.

SPICELPI.IN and the line per-unit-length **L** and **C** matrices from PUL.DAT and constructs the subcircuit model which, is output to the file SPICELPI. OUT. The source and load circuitry are added to this to give a complete SPICE circuit.

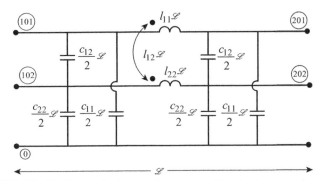

FIGURE 6.7. The Lumped-pi approximate model for the three-conductor line.

For the previous example shown in Fig. 6.4, the SPICELPI.IN file is

```
3               =Total Number of Conductors (N+1)
0.254           =Total Line Length (m)
```

and

The PCB.IN data are

```
3               =TOTAL NUMBER OF LANDS
30.             =NUMBER OF CONDUCTOR SUBSECTIONS
3               =NUMBER OF REFERENCE LAND
15.0            =LAND WIDTH (mils)
45.0            =EDGE-TO-EDGE LAND SEPARATION (mils)
47.0            =BOARD THICKNESS (mils)
4.7             =BOARD RELATIVE DIELECTRIC CONSTANT
```

The output data for L and C are in PUL.DAT:

```
1    1    1.38314E-06      =L ( 1, 1 )
1    2    6.91570E-07      =L ( 1, 2 )
2    2    1.10706E-06      =L ( 2, 2 )
1    1    2.96949E-11      =C ( 1, 1 )
1    2   -2.02619E-11      =C ( 1, 2 )
2    2    4.05238E-11      =C ( 2, 2 )
1    1    1.16983E-11      =C0 ( 1, 1 )
1    2   -7.30777E-12      =C0 ( 1, 2 )
2    2    1.46155E-11      =C0 ( 2, 2 )
```

```
NUMBER OF LANDS=    3
NUMBER OF DIVISIONS PER LAND=   30
REFERENCE LAND=    3
LAND WIDTH (mils)= 1.500E+01
EDGE-TO-EDGE SEPARATION (mils)= 4.500E+01
BOARD THICKNESS (mils)= 4.700E+01
RELATIVE DIELECTRIC CONSTANT= 4.700E+00
```

The SPICE-generated subcircuit model is output to the file SPICELPI. OUT:

```
*SUBCIRCUIT MODEL OF A MULTICONDUCTOR TRANSMISSION
+ LINE*
*LUMPED-PI APPROXIMATE STRUCTURE*
*       TOTAL NUMBER OF CONDUCTORS= 3
*       TOTAL LINE LENGTH (METERS)= 2.54000E-01
*       L(  1,  1)= 1.38314E-06
*       L(  1,  2)= 6.91570E-07
*       L(  2,  2)= 1.10706E-06
*       C(  1,  1)= 2.96949E-11
*       C(  1,  2)=-2.02619E-11
*       C(  2,  2)= 4.05238E-11
.SUBCKT LPI
+ 101
+ 102
+ 201
+ 202
L101   101   201      3.51318E-07
C101   101   0      1.19799E-12
C201   201   0      1.19799E-12
L102   102   202      2.81193E-07
C102   102   0      2.57326E-12
C202   202   0      2.57326E-12
K101   L101   L102      5.58879E-01
CM101   101   102      2.57326E-12
CM201   201   202      2.57326E-12
.ENDS       LPI
```

The terminations are added to this to give the SPICE.IN complete file

```
SPICE LUMPED PI MODEL
VS 1 0 PULSE(0 1 0 6.25N 6.25N 43.75N 100N
*VS 1 0 AC 1 0
RS 1 S 50
RL L 0 50
RNE NE 0 50
RFE FE 0 50
XLPI S NE L FE LPI
.TRAN 0.1 100N 0 0.1N
*.AC DEC 10K 1000MEG
.PROBE
*SUBCIRCUIT MODEL OF A MULTICONDUCTOR TRANSMISSION
+ LINE*
*LUMPED-PI APPROXIMATE STRUCTURE*
*          TOTAL NUMBER OF CONDUCTORS= 3
*          TOTAL LINE LENGTH (METERS)=2.54000E-01
*          L( 1, 1)=1.38314E-06
*          L( 1, 2)=6.91570E-07
*          L( 2, 2)=1.10706E-06
*          C( 1, 1)=2.96949E-11
*          C( 1, 2)=-2.02619E-11
*          C( 2, 2)=4.05238E-11
.SUBCKT LPI
+ 101
+ 102
+ 201
+ 202
L101    101    201    3.51318E-07
C101    101    0      1.19799E-12
C201    201    0      1.19799E-12
L102    102    202    2.81193E-07
C102    102    0      2.57326E-12
C202    202    0      2.57326E-12
K101    L101   L102   5.58879E-01
CM101   101    102    2.57326E-12
CM201   201    202    2.57326E-12
.ENDS       LPI
.END
```

The comparison of the predictions of the near- and far-end crosstalk voltages from the MTL subcircuit model and from the lumped-pi subcircuit model are given in Fig. 6.8.

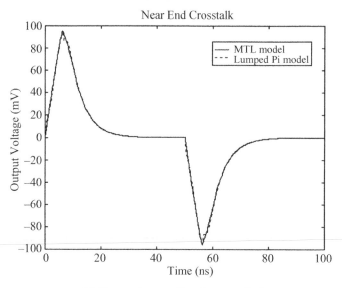

(a) Near end crosstalk voltage predictions.

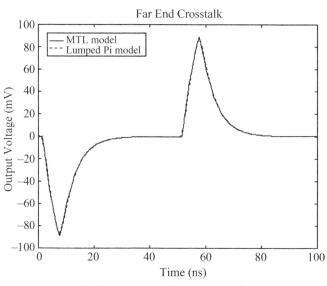

(b) Far end crosstalk voltage predictions.

FIGURE 6.8. Comparison of the predictions of the the near- and far-end crosstalk voltages of the SPICE subcircuit model and the lumped-pi subcircuit model for the problem of Fig. 6.4.

The predictions of the magnitudes of the frequency-domain transfer functions computed by the MTL subcircuit and lumped-pi models are compared in Fig. 6.9. The line is electrically short at around 70 MHz. The bandwidth of the source voltage waveform is approximately

$$\mathrm{BW} = \frac{1}{\tau_r} = \frac{1}{6.25\,\mathrm{ns}} = 160\,\mathrm{MHz}$$

Hence it is not surprising that the lumped-Pi model gives a reasonable prediction of the exact MTL model results.

6.4   A PRACTICAL CROSSTALK PROBLEM

Finally, we consider a common crosstalk problem found in digital circuits and shown in Fig. 6.10. It consists of two parallel CMOS data transmission lines. The lines are two parallel transmission lines 2 in in total length. The conductors are of width 10 mils and are separated 10 mils edge to edge as shown in Fig. 6.10(b). The lands lie on top of a 47 mil glass epoxy (FR4) substrate having a ground plane on the other side which serves as the reference conductor for both lines (a coupled microstrip line). Both lines are driven by CMOS drivers and receivers. The CMOS drivers are characterized by low-impedance (10-Ω) resistors at the source or near ends, and are characterized at the load or far ends by 5-pF capacitances, representing the input to the CMOS receivers. These are typcal CMOS output and input equivalent circuits. The task here is to compute the far-end crosstalk voltage, V_{FE}. Note that this problem can be visualized as a single-input, $V_S(t)$, single-output, $V_{FE}(t)$, system. The driving source on the generator line, $V_S(t)$, shown in Fig. 6.10(c) is a 100-MHz clock or data pulse train having a 50% duty cycle and rise and fall times of 1 ns. Hence the BW is BW = $1/\tau_r$ = 1 GHz.

First compute the per-unit-length parameters using MSTRP.FOR (MSTRP. EXE). The input file is MSTRP.IN:

```
2           =TOTAL NUMBER OF LANDS (EXCLUSIVE OF GND PLANE)
50          =NUMBER OF CONDUCTOR SUBSECTIONS
10.         =LAND WIDTH (mils)
10.         =EDGE-TO-EDGE LAND SEPARATION (mils)
47.         =BOARD THICKNESS (mils)
4.7         =BOARD RELATIVE DIELECTRIC CONSTANT.
```

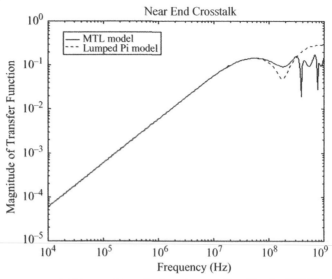

(a) Near end transfer functions predicted by the SPICE
and Lumped Pi subcircuit models.

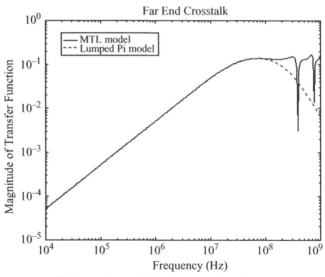

(b) Far end transfer functions predicted by the SPICE
and Lumped Pi subcircuit models.

FIGURE 6.9. Comparison of the near-end and far-end transfer functions predicted by the
SPICE subcircuit models and the lumped-Pi subcircuit models for the problem of Fig. 6.4.

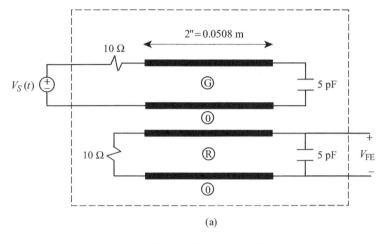

(a)

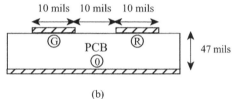

(b)

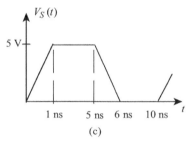

(c)

FIGURE 6.10

Then execute MSTRP.FOR (MSTRP.EXE) to yield the per-unit-length line parameters in PUL.DAT:

```
1          1          7.20737E-07          =L (  1,  1)
1          2          3.21065E-07          =L (  1,  2)
2          2          7.20738E-07          =L (  2,  2)
1          1          5.72329E-11          =C (  1,  1)
1          2          -2.24421E-11         =C (  1,  2)
2          2          5.72331E-11          =C (  2,  2)
1          1          1.92595E-11          =C0 (  1,  1)
1          2          -8.57948E-12         =C0 (  1,  2)
2          2          1.92595E-11          =C0 (  2,  2)
```

```
NUMBER OF LANDS=  2
NUMBER OF DIVISIONS PER LAND= 50
LAND WIDTH (mils)= 1.000E+01
EDGE-TO-EDGE SEPARATION (mils)= 1.000E+01
BOARD THICKNESS (mils)= 4.700E+01
RELATIVE DIELECTRIC CONSTANT= 4.700E+00
```

Next, prepare SPICEMTL.IN:

```
3                      =Total Number of Conductors (N+1)
0.0508                 =Total Line Length (m)
```

Submit SPICEMTL.FOR (SPICEMTL.EXE) giving the SPICE subcircuit for Fig. 6.2 in SPICEMTL.OUT. Adding the terminations to this gives the complete SPICE model:

```
PCB XTALK TWO 10mil LANDS SEPARATED BY 10mils ON 47mil
+ PCB
VS 1 0 PULSE (0 5 0 1N 1N 4N 10N )
*VS 1 0 AC 1 0
RS 1 2 10
CL 3 0 5P
RNE 4 0 10
CFE 5 0 5P
XMTL 2 4 3 5 MTL
.TRAN 0.01N 40N 0 0.01N
.PRINT TRAN V(1) V(5)
*.AC DEC 10000 10E6 1E10
*.PRINT AC VDB (5) VP(5)
.PROBE
*EXACT TRANSMISSION LINE MODEL
*SUBCIRCUIT MODEL OF A MULTICONDUCTOR TRANSMISSION
+ LINE*
*      NUMBER OF CONDUCTORS=  3
*      TOTAL LINE LENGTH (METERS)= 5.08000E-02
*      L ( 1,  1) = 7.20737E-07
*      L ( 1,  2) = 3.21065E-07
*      L ( 2,  2) = 7.20738E-07
*      C ( 1,  1) = 5.72332E-11
*      C ( 1,  2) = -2.24419E-11
*      C ( 2,  2) = 5.72330E-11
.SUBCKT MTL
+ 101
+ 102
```

```
+ 201
+ 202
V101  101   301
EC101  301  0     POLY ( 2 )
+   (501,0)
+   (502,0)
+ 0
+     7.070959E-01
+     7.071106E-01
FC101  0   501  POLY ( 2 )
+ V101
+ V102
+ 0
+     7.070959E-01
+    -7.071176E-01
V201   201   401
EC201 401 0     POLY ( 2 )
+   (601,0)
+   (602,0)
+ 0
+     7.070959E-01
+     7.071106E-01
FC201   0   601  POLY ( 2 )
+ V201
+ V202
+ 0
+     7.070959E-01
+    -7.071176E-01
V102   102    302
EC102   302    0  POLY ( 2 )
+ (501,0)
+ (502,0)
+ 0
+    -7.071176E-01
+     7.071030E-01
FC102 0 502       POLY ( 2 )
+ V101
+ V102
+ 0
+     7.071106E-01
+     7.071030E-01
V202   202   402
```

```
EC202   402   0   POLY(2)
+ (601,0)
+ (602,0)
+ 0
+   -7.071176E-01
+    7.071030E-01
FC202   0   602    POLY(2)
+ V201
+ V202
+ 0
+    7.071106E-01
+    7.071030E-01
T101   501  0  601  0   Z0=7.082574E+01  TD=2.866664E-10
T102   502  0  602  0   Z0=1.730446E+02  TD=3.058378E-10
.ENDS      MTL
.END
```

Since there are two lines, there are two mode velocities. SPICE gives the one-way time delays as $T_{D1} = 0.28666$ ns and $T_{D2} = 0.30584$ ns. Hence, the mode lines are one wavelength long at 3.5 and 3.3 GHz. So the line is electrically short at around 350 or 330 MHz. The characteristic impedances of the two modes are 70.8 and 173 Ω.

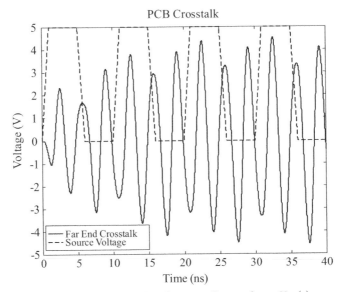

FIGURE 6.11. Time-domain crosstalk waveform, $V_{FE}(t)$.

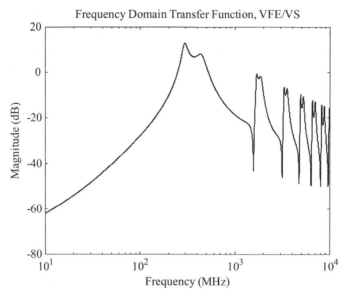

FIGURE 6.12. The magnitude of the frequency-domain crosstalk transfer function, $\hat{V}_{FE}/\hat{V}_S$.

Figure 6.11 compares the source voltage waveform, $V_S(t)$, and the calculated far-end crosstalk waveform of the nearby line, $V_{FE}(t)$, at the input to the second CMOS receiver. *Observe that these results show an amazing and unexpected crosstalk level of almost twice as large as the source voltage even*

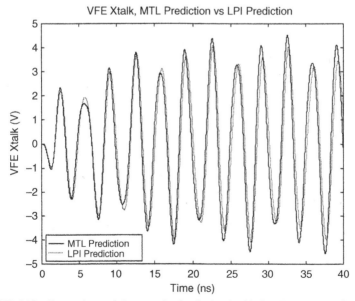

FIGURE 6.13. Comparison of the magnitude obtained with SPICEMTL and SPICELPI.

though the two circuits are NOT "hardwired"! This shows rather dramatically that crosstalk can be a significant interference problem.

Figure 6.12 shows the magnitude of the frequency-domain crosstalk transfer function. Note that there is a resonance at around 300 MHz. This resonance is due to the 5-pF load capacitors of the CMOS inputs. A lumped-pi approximate model shows that the far-end parameters resonate along with the 5-pF load to give

$$
f_0 = \frac{1}{2\pi\sqrt{L_R\left(\dfrac{C_R + C_m}{2} + 5\,\text{pF}\right)}}
$$

$$
= 327.4\,\text{MHz}
$$

where the total inductances and capacitances are $L_R = l_R \times \mathscr{L} = 36.6\,\text{nH}$ and $C_R + C_m = (c_R + c_m) \times \mathscr{L} = 2.91\,\text{pF}$. These give a resonance frequency of 327.4 MHz, which matches rather well with the first resonance in Fig. 6.12. Above that frequency we see the effects of the transmission line begin to come into play. The line is one wavelength at around 3.5 GHz. Points where the line length is a multiple of a half-wavelength are clearly shown and are repeating repetitively, as expected.

Figure 6.13 shows the excellent correlation between the transmission line model and the lumped-Pi model.

APPENDIX

A BRIEF TUTORIAL ON USING PSPICE

This is a brief summary of the SPICE simulation program with integrated-circuit emphasis, or its personal computer version PSPICE, electric circuit analysis program. The original SPICE computer program was developed to analyze complex electric circuits, particularly integrated circuits. It was developed at the University of California at Berkeley in the late 1960s. Since it was developed under U.S. government funding, it is not proprietary and can be freely copied, used, and distributed. This was written for use on large mainframe computers of the time. In the early 1980s the MicroSim Corporation developed a personal computer version of SPICE called PSPICE. A number of important modifications were made, particularly in the plotting of data via the .PROBE function. Since then a number of commercial firms have modified and developed their own PC versions. But essentially the core engine is that of the original SPICE code. The MicroSim version of PSPICE, version 8, was acquired by Cadence Design Systems. Theirs is version 10.0, called OrCAD Capture. A Windows-based version is available free from www.cadence.com. The OrCAD Capture program was originally called Schematic in the MicroSim version and contains a number of enhancements. A number of books [1–5] detail the use of SPICE and PSPICE. Both the MicroSim version 8 and the OrCAD version 10 are contained in a CD at the end of other textbooks [6].

Transmission Lines in Digital Systems for EMC Practitioners, First Edition.
Clayton R. Paul.
© 2012 John Wiley & Sons, Inc. Published 2012 by John Wiley & Sons, Inc.

There are two methods of entering and executing a PSPICE program. The first method is the *Direct Method,* described here, where one enters the program code using an ASCII text editor (supplied with PSPICE). *Note:* SPICE and PSPICE make no distinction between lowercase and uppercase letters. Then this text file is run using the PSpice A/D section of the program and the output is examined with the text editor. The second method is the *Schematic Method* (now called Capture), where the user "draws" the circuit diagram directly on the screen and then executes that program. The Direct Method is generally the more rapid method of solving relatively simple problems. The Schematic (Capture), method has the advantage of visually seeing whether the circuit components are connected as intended for more complex circuits, but for most simple problems is a bit more time consuming to set up than the Direct Method since numerous windows and drop-down menus must be navigated in the Schematic or Capture method.

Once the PSPICE program has been installed on your computer, the following is a description of how you can input your program, run it, and examine the output. The various selections are underlined. Although there are several ways of doing this, the simplest is to use the Design Manager. To load this, you click or select the following in this sequence.

1. Start
2. Programs
3. MicroSim Eval 8
4. Design Manager

The Direct Method is simply to type in the PSPICE program using the TextEdit feature. To enter this and prepare the program, we select the following in this sequence.

1. Select TextEdit (lower button on the vertical toolbar on the left).
2. Type the program.
3. Save the program as XXX.cir or XXX.in and close it.
4. Select PSpice A/D (second button on the vertical toolbar on the left).
5. Click on File, Open and select the file stored previously. The program will run automatically and the output will be stored in file XXX.out.
6. Click on File, Run Probe to plot waveforms or File, Examine Output to examine the printed output.
7. Alternatively, you could recall the TextEdit progam and select File, Open, XXX.out to examine the printed output, which is self-explanatory.

CREATING THE SPICE OR PSPICE PROGRAM

SPICE and PSPICE write the node-voltage equations of an electric circuit [1]. One node, the reference node for the node voltages, is designated the zero (0) node. All circuits must contain a zero node. The other nodes in the circuit are labeled with numbers or letters. For example, a node may be labeled 23 or it may be labeled FRED. The voltages with respect to the reference node are positive at the node and denoted as V(N1), V(N2), and so on, as shown in Fig. A.1.

The general structure of any SPICE or PSPICE program is as follows:

1. Title

2. { Circuits Description

3. Execution Statements
4. OutputStatements
5. .END

The first line of the SPICE program is the Title and is not processed by SPICE. It is simply written on the output and on any plots. A comment line is started with an asterisk (*) and is also not processed by the program. A line may be continued with a plus sign (+) at the beginning of the following line. The next set of lines, Circuit Description, describes the circuit elements and their values and tells SPICE how they are connected together to form the circuit. The next set of lines are the Execution Statements that tell SPICE what type of analysis is to be run: dc analysis (.DC), sinusoidal steady-state

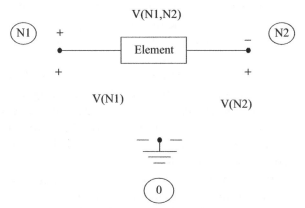

FIGURE A.1 Node voltage and element voltage notation in the SPICE (PSPICE) circuit analysis program.

or phasor analysis (.AC), or the full time-domain analysis consisting of the transient and steady-state solution (.TRAN). The next set of statements, Output Statements, tell SPICE what outputs are desired. The results can be printed to a file with the .PRINT statement or can be plotted with the .PROBE feature. Finally, all programs must end with the .END statement. Actually, the items 2 to 4 above can appear in any order in the program, but the program must begin with a Title statement and end with the .END statement.

CIRCUIT DESCRIPTION

The basic elements and their SPICE descriptions are shown in Fig. A.2. Figure A.2(a) shows the independent voltage source. It is named starting with the letter V and then any other letters. For example, a voltage source might be called VFRED. It is connected between nodes N1 and N2. It is very important to note that *the source is assumed positive at the first-named node*. The current through the voltage source is designated as I(VXXX) and is assumed to flow from the first-named node to the last-named node. The source type can be either dc, for which we append the term DC *magnitude*, or a sinusoid, to which we append the term AC *magnitude phase* (*degrees*). A time-domain waveform is described by several functions which we describe later, and these descriptions are appended (without the word TRAN). The independent current source is shown in Fig. A.2(b). Its name starts with the letter I followed by any other letters. For example, a current source might be designated as ISAD. *The current of the source is assumed to flow from the first-named node to the last-named node*. The types of source waveforms are the same as for the voltage source.

The resistor is shown in Fig. A.2(c) and its name starts with the letter R (e.g., RHAPPY). The current through the resistor is designated as I(RXXX) and is assumed to flow from the first-named node to the last-named node. SPICE does not allow elements with zero values. Hence a resistor whose value is $0\,\Omega$ (a short circuit) may be represented as having a value of 1E-8 or any other suitably small value. Similarly, an open circuit may be designated as a resistor having a value of 1E8 or any other suitably large value. Every node must have at least two elements connected to it. Also, SPICE requires that every node must have a dc path to ground [the zero (0) node]. Placing a large resistor (e.g., $1\,\text{M}\Omega$) between such nodes fixes this problem.

The inductor is shown in Fig. A.2(d) and is designated by the letter L (e.g., LTOM). The current through the inductor as well as the initial inductor current at $t = 0^{+}$, I(0), is assumed to flow from the first-named

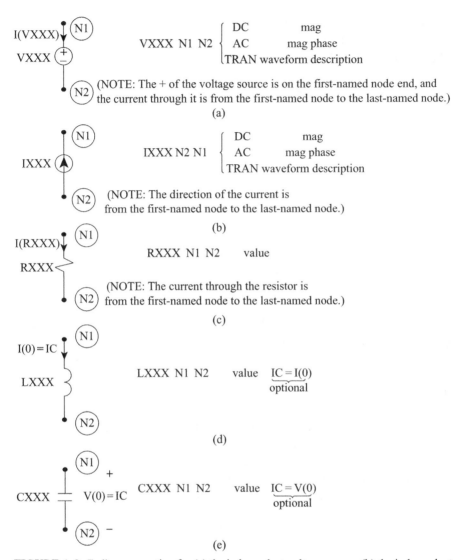

FIGURE A.2 Coding convention for (a) the independent voltage source, (b) the independent current source, (c) the resistor, (d) the inductor, and (e) the capacitor.

node to the last-named node. The initial condition can be specified at the end of the statement with $IC = I(0)$. The capacitor is shown in Fig. A.2(e) and is designated by the letter C (e.g., CME). The initial voltage across the capacitor at $t = 0^+$, $V(0)$, can be specified at the end of the statement with $IC = V(0)$, and this voltage is assumed to be positive at the first-named node.

All numerical values can be specified in powers of 10 and written in exponential format (e.g., $2 \times 10^{-5} = 2E - 5$), or by using standard multipliers using standard engineering notation:

Multiplier	SPICE Symbol
10^9 (giga)	G
10^6 (mega)	MEG
10^3 (kilo)	K
10^{-3} (milli)	M
10^{-6} (micro)	U
10^{-9} (nano)	N
10^{-12} (pico)	P

For example, 1 megohm is written as 1MEG, 1 kilohm is written as 1K, 3 millihenries is written as 3M, 5 microfarads is written as 5U, 2 nanohenries is written as 2N, and 7 picofarads is written as 7P. A 3-F capacitor should not be written as 3F, since F stands for femto $= 10^{-15}$.

The four types of controlled sources, G, E, F, and H, are shown in Fig. A.3 along with their descriptions. The polarities of voltage and the currents through the elements conform to the previous rules governing these in terms of the first- and last-named nodes on their description statements. For a current-controlled source, F or H, the controlling current must be through an independent voltage source. Often we insert a 0-V source to sample the current. Some recent versions of PSPICE allow specification of the current through any element as a controlling current. But it is always a simple matter to insert a zero-volt voltage source.

Figure A.4 shows how to specify mutual inductance. First the self-inductances that are coupled are specified as before. The mutual inductance is specified in terms of its coupling coefficient:

$$k = \frac{M}{\sqrt{L_1 L_2}}$$

To keep the polarities correct, define the self-inductances so that the dots are on the first-named nodes when the two inductors are defined; otherwise, a negative coupling coefficient may need to be specified.

Figure A.5 shows the last important element, the transmission line (lossless), which is used extensively in signal integrity analyses. This gives the *exact solution* of the transmission-line equations for a *lossless* line. There are many ways to specify the important parameters for the (lossless) line, but the one shown in the figure is the most widely used; specify the characteristic impedance of the line and the line's one-way time delay. Alternatively, you can create a lumped RLCG *approximate* model of a *lossy* line using

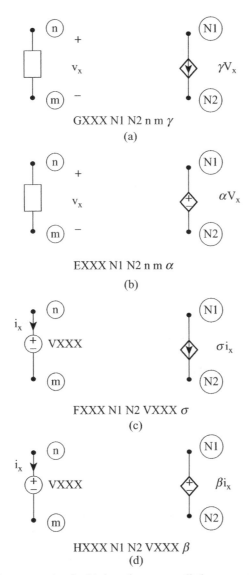

GXXX N1 N2 n m γ

(a)

EXXX N1 N2 n m α

(b)

FXXX N1 N2 VXXX σ

(c)

HXXX N1 N2 VXXX β

(d)

FIGURE A.3 Coding convention for (a) the voltage-controlled current source, (b) the voltage-controlled voltage source, (c) the current-controlled current source, and (d) the current-controlled voltage source.

lumped-circuit elements [6]. But the per-unit-length parameters of resistance R and conductance G in that model must be constants, although these are, in reality, frequency dependent, which is not readily handled in an exact solution of the transmission-line equations [6].

Figure A.6 shows how to specify the important time-domain waveforms. Figure A.6(a) shows the PWL (piecewise-linear) waveform, where straight

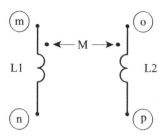

L1 m n   value
L2 o p value
KXXX L1 L2 coefficient coupling

$$\text{coupling coefficient } \frac{M}{\sqrt{L1\ L2}}$$

FIGURE A.4 Coding convention for mutual inductance between two coupled inductors.

lines are drawn between pairs of points that are specified by their time location and their value. Observe that the function holds the last specified value, V4 in the figure. Figure A.6(b) shows the periodic pulse waveform, PULSE, that is used to specify periodic clock or other timing waveforms. The function specifies a trapezoidal waveform that repeats periodically with period PER (the reciprocal is the fundamental frequency of the waveform). Note that the pulse width, PW, is not specified between the 50% points of the pulse, as is the usual convention.

The sinusoidal function is specified by

$$\text{SIN(Vo Va [[Freq [[Td [[Df [[Phase]]]]]]])}$$

which gives the waveform

$$x(t) = \text{Vo} + \text{Va}\sin\left(2\pi\left(\text{Freq(time} - \text{Td)} + \frac{\text{Phase}}{360}\right)\right)e^{-(\text{time}-\text{Td})\text{Df}}$$

Brackets around items signify that they are optional. Hence to specify the general sinusoidal waveform

$$x(t) = A\sin(n\omega_0 t + \theta)$$

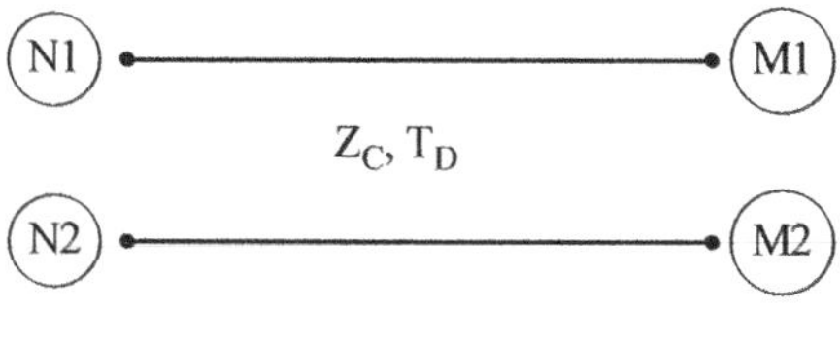

N1 •————————————• M1

$$Z_C, T_D$$

N2 •————————————• M2

TXXX N1 N2 M1 M2   Z0 = Z_C TD = T_D

FIGURE A.5 Coding convention for the two-conductor lossless transmission line.

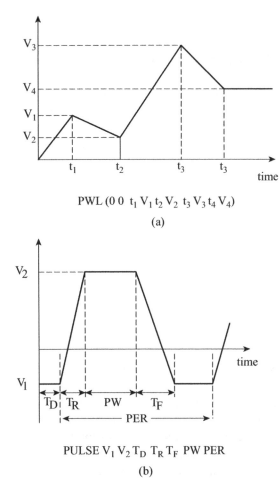

$$PWL\ (0\ 0\ t_1\ V_1\ t_2\ V_2\ t_3\ V_3\ t_4\ V_4)$$

(a)

$$PULSE\ V_1\ V_2\ T_D\ T_R\ T_F\ PW\ PER$$

(b)

FIGURE A.6 Coding convention for the important source waveforms: (a) the piecewise-linear waveform and (b) the pulse source waveform (periodic).

where $\omega_0 = 2\pi f_0$ and $f_0 = 1/PER$ is the frequency of the sinusoid, we would write

$$SIN(0\ A\ nf_0\ 0\ 0\ \theta)$$

EXECUTION STATEMENTS

There are three types of solutions: dc, sinusoidal steady state or phasor, and the full time-domain solution (so-called *transient*, although it contains both the transient and the steady-state parts of the solution).

The dc solution is specified by

.DC V, IXXX start_value end_value increment

where V,IXXX is the name of a dc voltage or current source in the circuit whose value is to swept. For example, to sweep the value of a dc voltage source VFRED from 1 V to 10 V in increments of 2 V and solve the circuit for each of these source values, we would write

.DC VFRED 1 10 2

If no sweeping of any source is desired, simply choose one dc source in the circuit and iterate its value from the actual value to the actual value and use any nonzero increment. For example, a 5 V DC source would be specified as

.DC VFRED 5 5 1

The sinusoidal steady-state or phasor solution is specified by

.AC {LIN,DEC,OCT} points start_frequency end_frequency

LIN denotes a linear frequency sweep from *start_frequency* to *end_frequency* and *points* is the total number of frequency points. DEC denotes a log sweep of the frequency where the frequency is swept logarithmically from the *start_frequency* to the *end_frequency* and *points* is the number of frequency points per decade. OCT is a log sweep by octaves where *points* is the number of frequency points per octave.

The time-domain solution is obtained by specifying

.TRAN print_ step end_ time [no_ print_time] [step_ceiling] [UIC]

SPICE solves the time-domain differential equations of the circuit by discretizing the time variable into increments of Δt and solving the equations in a bootstrapping manner. The differential equations of the circuit are first solved at $t = 0$. Then that solution is used to give the solution at $t = \Delta t$. These prior time solutions are then used to give the solution at $t = 2\Delta t$, and so on. The first item, *print_step*, governs when an output is requested. Suppose that the discretization used in the solution is every 2 ms. We might not want to see (in the output generated by the .PRINT statement) an output at every 2 ms but only every 5 ms. Hence we might set the *print_step* time as 5M. The *end_time* is the final time for which the solution is obtained.

The remaining parameters are optional. The analysis always starts at $t=0$. But we may not wish to see a printout of the solution (in the output generated by the .PRINT statement) until after some time has elapsed. If so, we would set the *no_print_time* to that starting time. SPICE and PSPICE have a very sophisticated algorithm for determining the minimum step size, Δt, for discretization of the differential equations in order to get a valid solution. The default maximum step size is *end_time*/50. However, there are many cases where we want the step size to be smaller than what SPICE would allow, to increase the accuracy of the solution or to increase the resolution of the solution waveforms. This is frequently the case when we use SPICE in the analysis of transmission lines where abrupt waveform transitions are occurring at every one-way time delay. The *step_ceiling* is the maximum time step size, $\Delta t|_{max}$, that will be used in discretizing the differential equations, as described previously. Although this gives longer run times, there are cases where we need to do this to generate the required accuracy. The last item, *UIC*, means that SPICE is to use the initial capacitor voltage or inductor current specified on the element specifications with the $IC=$ command. In a transient analysis, SPICE will compute the initial conditions at $t = 0^+$. If some other initial conditions are required, we should set these on the capacitor or inductor specifications with the $IC=$ command and specify *UIC* on the .TRAN statement. For example,

$$.TRAN\ 0.5N\ 20N\ 0\ 0.01N$$

would command SPICE to do a full time-domain (transient plus steady state) analysis for times from 0 to 20 ns, print out a solution at every 0.5 ns, start printing to the output file at $t=0$, and use a time discretization time step no larger than 0.01 ns.

OUTPUT STATEMENTS

The output statements are either for printing to a file with the .PRINT statement or producing a plotted graph of any waveform with the .PROBE statement. The .PRINT statement has three forms, depending on the type of analysis being run. For a DC analysis,

$$.PRINT\ DC\ V(X)\ I(R)$$

prints the dc solution for the voltage of node X with respect to the reference node, and I(R) prints the dc solution for current through resistor R (defined from the first-named node to the last-named node on the specification statement for resistor R). For a sinusoidal steady-state analysis (phasor solution),

.PRINT AC VM(NI) VP(NI) IM(RFRED) IP(RFRED)

prints the magnitude and phase of node voltages and currents where the magnitude and phase of the node voltage at node NI are VM(NI) and VP(NI), respectively. For the currents through a resistor RFRED, the magnitude is IM (RFRED) and the phase is IP(RFRED). For the time-domain or transient analysis, the print statement

.PRINT TRAN V(NI) I(RFRED)

prints the solutions at all solution time points (specified on the .TRAN line) for the voltage at node NI with respect to the reference node, and the current through resistor RFRED (defined from the first-named node to the last-named node on the specification statement for resistor RFRED).

In addition, the .FOUR statement computes the expansion coefficients for the Fourier series (magnitude and phase), $c_n \angle c_n$:

.FOUR f_0 [output_variable(s)]

The .FOUR command can only be used in a .TRAN analysis. The fundamental frequency of the *periodic* waveform to be analyzed is denoted as $f_0 = 1/T$, where T is the period of the waveform. The *output_variable(s)* are the desired voltage or current waveforms to be analyzed [e.g., V(2), I(R1)]. The phase results are with reference to a sine form of the series:

$$x(t) = c_0 + \sum_{n=1}^{\infty} c_n \sin(n\omega_0 t + \angle c_n)$$

There is an important consideration in using the .FOUR command. The portion of the waveform that is analyzed to give the Fourier expansion coefficients is the last portion of the solution time of length one period, $1/f_0 = T$. In other words, SPICE determines the coefficients from the waveform between *end_time* $- [1/f_0]$ and *end_time*. Hence, *end_time* on the .TRAN command should be at least one period long. In situations where the solution has a transient portion at the beginning of the solution interval and we want to determine the Fourier coefficients for the steady-state solution, we would run the analysis for several periods to ensure that the solution has gotten into steady state. For example, consider an input signal that is periodic with a period of 2 ns or a fundamental frequency of 500 MHz. An output voltage at node 4 would also have this periodicity but would have a transient period of at least five time constants. Suppose that the maximum time constant of the circuit is 4 ns. Then we would set the end time to 20 ns or more to get into the steady-state region of the solution at the end of the solution. The following commands would be used to obtain the Fourier coefficients of the steady-state response of the node voltage at node 4:

.TRAN 0.1N 20N
.FOUR 500MEG V(4)

This would compute the solution for the voltage waveform at node 4 from $t = 0$ to $t = 20$ ns. Since the period (the inverse of 500 MHz) is specified as 2 ns, the portion of the waveform from 18 to 20 ns would be used to compute the Fourier coefficients for the waveform. If we wanted to compute the Fourier coefficients for the initial part of the waveform including the transient, we would specify

.TRAN 0.1N 2N

which would run for only one period. There is a FFT button on the toolbar that can compute the fast Fourier transform of the waveform. Essentially this valuable feature can turn PSPICE into a "poor man's spectrum analyzer."

All printed output is directed to a file named XXXX.OUT if the input file is named XXXX.IN or XXXX.CIR. Plotting waveforms is the greatest enhancement of PSPICE over the original SPICE. This is invoked simply by placing the .PROBE statement in the program list. No additional parameters are required. PSPICE stores all variables at all solution time points and waits for the user to specify which to plot.

EXAMPLES

In this brief tutorial we have shown the basic commands that one can use to solve the vast majority of electric circuit analysis problems. Essentially, PSPICE performs the tedious and laborious solutions of the lumped-circuit models of the problem. Solving these circuits by hand would not only be time consuming but would involve so many errors as to make the solutions obtained useless. To simplify the learning we have conscientiously tried to minimize the detail and purposely not shown all the possible options. However, there are a myriad of options that can simplify many computations, and the reader should consult the references.

EXAMPLE 1

Use PSPICE to compute the voltage V_{out} and the current i in the circuit of Fig. A.7.

Solution The PSPICE coding diagram with nodes numbered is shown in Fig. A.7. Zero-volt voltage sources are inserted to sample the currents i_x and i. The PSPICE program is

```
EXAMPLE 1
VS 1 0 DC 5
R1 1 2 500
R2 2 3 1K
R3 3 4 2K
VTEST1 4 0 DC 0
HSOURCE 3 5 VTEST1 500
R4 5 6 500
VTEST2 6 0 DC 0
.DC VS 5 5 1
*THE CURRENT I IS I(VTEST2) AND THE VOLTAGE VOUT IS V(3)
+OR V(3,4)
.PRINT DC V(3) I(VTEST2)
.END
```

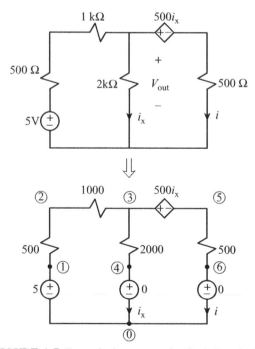

FIGURE A.7 Example 1, an example of a DC analysis.

EXAMPLE 2   **259**

The result is $i = I(VTEST2) = 1.875E-3$ and the voltage $Vout = V(3) = 1.250E0$.

EXAMPLE 2

Use PSPICE to plot the frequency response of the bandpass filter shown in Fig. A.8(a).

Solution The nodes are numbered on the circuit diagram and the PSPICE program is

```
EXAMPLE 2
VS 1 0 AC 1 0
RES 3 0 100
LIND 1 2 159U
CAP 2 3 1.6P
```

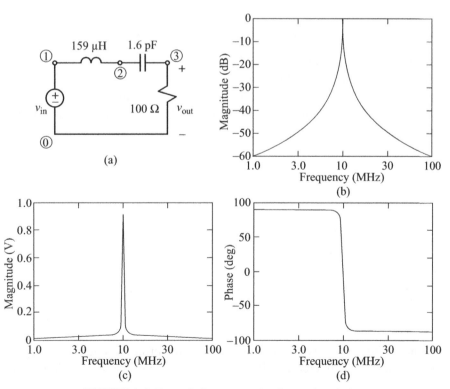

FIGURE A.8 Example 2, an example of an AC analysis.

```
.AC DEC 50 1MEG 100MEG
.PROBE
*THE MAGNITUDE OF THE OUTPUT IS VM(3) AND THE PHASE IS VP(3)
.END
```

The magnitude of the voltage is plotted in Fig. A.8(b) in decibels using VDB(3), which means that

$$VDB(3) = 20 \ \log_{10} VM(3)$$

Figure A.8(c) shows what we get if we request VM(3): The data are highly compressed outside the bandpass region. The phase (in degrees) is plotted in Fig. A.8(d). The resonant frequency is 10 MHz. The phase is $+90°$ below the resonant frequency, due to the dominance of the capacitor in this range and is $-90°$ above the resonant frequency due to the dominance of the inductor in this range. This bears out the important behavior of a series resonant circuit.

EXAMPLE 3

Use PSPICE to plot the inductor current for $t > 0$ in the circuit of Fig. A.9(a).

Solution The circuit immediately before the switch opens (i.e., at $t = 0^-$) is shown in Fig. A.9(b), from which we compute the initial voltage of the capacitor as 4 V and the initial current of the inductor as 2 mA. The PSPICE diagram with nodes numbered is shown in Fig. A.9(c), and the PSPICE program is

```
EXAMPLE 3
IS 0 1 DC 10M
R 1 2 2K
VTEST 2 3
L 3 0 10M IC=2M
C 1 0 100P IC=4
.TRAN .1U 50U 0 .05U UIC
*THE INDUCTOR CURRENT IS I(VTEST) OR I(L)
.PROBE
.END
```

We have chosen to solve the circuit out to 50 μs, print the solution in steps of 0.1 μs, and have directed PSPICE to use a solution time step no larger that

EXAMPLE 3   **261**

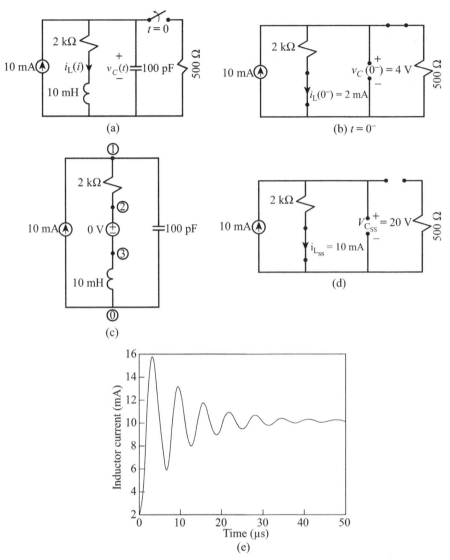

FIGURE A.9 Example 3, an example of a TRAN analysis.

0.05 μs, as well as to use the initial conditions given for the inductor and capacitor. The result is plotted using PROBE in Fig. A.9(e). The result starts at 2 mA, the initial inductor current, and eventually converges to the steady-state value of 10 mA, which can be confirmed by replacing the inductor with a short circuit and the capacitor with an open circuit in the $t > 0$ circuit, as shown in Fig. A.9(d).

EXAMPLE 4

Figure A.10 shows an example where an interconnecting set of conductors (lands on a PCB) can cause severe logic errors, resulting in poor signal integrity. Two CMOS inverters (buffers) are connected by 2 in of lands ($\mathscr{L} = 2\,\text{in} = 5.08\,\text{cm}$) on a PCB. The output of the left inverter is shown as a Thévenin equivalent circuit having a low source resistance of $10\,\Omega$. This is fairly typical of CMOS devices except that the output resistance is somewhat

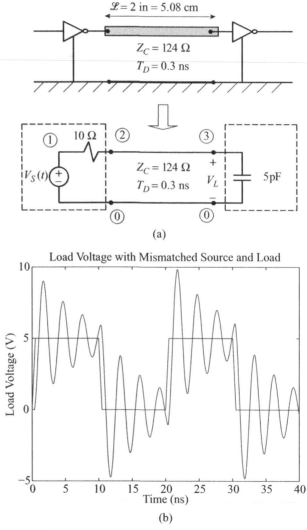

(a)

(b)

FIGURE A.10 Example 4, an example of a signal integrity analysis.

nonlinear. The load on the line is the input to the other CMOS inverter, which is represented as a 5-pF capacitor that is also typical of the input to CMOS devices. We are interested in determining the voltage at the output of the interconnect line, $V_L(t)$, which is the voltage at the input to the second CMOS inverter. The open-circuit voltage of the left inverter, $V_S(t)$, is a 5-V, 50-MHZ (period of 20 ns), 50% duty cycle, clock trapezoidal waveform having 0.5-ns rise and fall times. A 10-mil land lying on a glass-epoxy PCB of 47 mils thickness has a ground plane below it. For the cross-sectional dimensions of the line, the characteristic impedance is $Z_C = 124\,\Omega$, and the velocity of propagation is $v = 1.7 \times 10^8$ m/s. This gives a one-way time delay of

$$T_D = \frac{\mathscr{L}}{v} = \frac{0.0508\ \text{m}}{1.7 \times 10^8\ \text{m/s}} = 0.3\ \text{ns}$$

The response for the load voltage is computed with PSPICE using the exact (lossless) transmission-line model contained in PSPICE and is shown in Fig. A.10(b).

Solution The nodes are numbered on the circuit diagram and the PSPICE program is

```
EXAMPLE 4
VS 1 0 PULSE(0 5 0 0.5N 0.5N 9.5N 20N)
RS 1 2 10
T 2 0 3 0 Z0=124 TD=0.3N
CAP 3 0 5P
.TRAN 0.05N 40N 0 0.05N
.PROBE
.END
```

Typical thresholds for CMOS circuits are around halfway between the logic 1 and logic 0 levels, which in this case are 5 V and 0 V. Observe that there is severe "ringing" in the response (the input to the second inverter) and the response drops below the 2.5 V high level and rises above the 2.5 V low level, thereby producing false logic triggering. Hence signal integrity is not achieved here.

THE SUBCIRCUIT MODEL

SPICE (PSPICE) has a handy way of utilizing models of devices in several places in a SPICE program without having to redefine these models at every

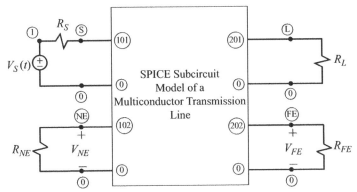

FIGURE A.11 Connection of a SPICE subcircuit model of a multiconductor transmission line (MTL) to its terminations.

place of usage. This is similar to the subroutine in FORTRAN. It is called the SUBCKT model. For example, suppose that we have developed an extensive model of, say, an op amp or a multiconductor transmission line (MTL) [6]. The subcircuit model might have, for example, four external nodes that we have named 101, 102, 201, 202, as illustrated in Fig. A.11 for a subcircuit model of a MTL.

The nodes internal to the model are unique to this model and have no resemblance to the nodes of the SPICE program into which this model is to be embedded (perhaps at several locations). However, the zero (0) or universal ground node is the only node that is common with the main program. The subcircuit model description is

.SUBCKT MTL 101 102 201 202

. . . . . .

. . . .

.ENDS MTL

where MTL is the name given to the subcircuit, and its external nodes are 101, 102, 201, and 202. The subcircuit model ends with .ENDS and the name of the subcircuit model, MTL. These node numbers are unique to the subcircuit, *but their ordering is important*. The call statement in the SPICE main program is formatted as

XMTL S NE L FE MTL

Hence the nodes of the subcircuit MTL are attached to the external nodes of the main SPICE model as $S = 101$, $NE = 102$, $L = 201$, $FE = 202$ to connect the subcircuit to the MTL terminations, as illustrated in Fig. A.11. The subcircuit model must end with

.ENDS MTL

and the main SPICE program must end with the statement

.END

REFERENCES

1. C. R. Paul, *Fundamentals of Electric Circuit Analysis*, Wiley, New York, 2001.
2. P. W. Tuinenga, *SPICE: A Guide to Simulation and Analysis Using PSPICE*, 3rd ed., Prentice Hall, Englewood Cliffs, NJ, 1995.
3. A. Vladimirescu, *The SPICE Book*, Wiley, New York, 1994.
4. R. Conant, *Engineering Circuit Analysis with PSpice and Probe*, McGraw-Hill, New York, 1993.
5. J. W. Nilsson and S. A. Riedel, *Introduction to PSpice Manual for Electric Circuits Using OrCad Release 9. 1*, 4th ed., Prentice Hall, Upper Saddle River, NJ, 2000.
6. C. R. Paul, *Introduction to Electromagnetic Compatibility*, 2nd ed., Wiley, Hoboken, NJ, 2006.

INDEX

Transmission Lines in Digital Systems for EMC Practitioners, First Edition.
Clayton R. Paul.
© 2012 John Wiley & Sons, Inc. Published 2012 by John Wiley & Sons, Inc.

Printed and bound by CPI Group (UK) Ltd, Croydon, CR0 4YY

27/10/2024

14580265-0005